Evolutionary conserved features of FG repeats that
allow the formation of hydrogel-based permeability
barriers with NPC-like properties

Evolutionary conserved features of FG repeats that allow the formation of hydrogel-based permeability barriers with NPC-like properties

Dissertation

in partial fulfillment of the requirements
for the degree *'Doctor rerum naturalium'*
in the Molecular Biology Program
at the Georg-August University Göttingen,
Faculty of Biology

Submitted by
Hermann Broder Schmidt

Born in
Itzehoe, Germany

Göttingen, September 2012

Bibliografische Information der Deutschen Nationalbibliothek
Die Deutsche Nationalbibliothek verzeichnet diese Publikation in der
Deutschen Nationalbibliografie; detaillierte bibliografische Daten sind im Internet
über http://dnb.d-nb.de abrufbar.
1. Aufl. - Göttingen : Cuvillier, 2013
   Zugl.: Göttingen, Univ., Diss., 2012

   978-3-95404-491-7

© CUVILLIER VERLAG, Göttingen 2013
   Nonnenstieg 8, 37075 Göttingen
   Telefon: 0551-54724-0
   Telefax: 0551-54724-21
   www.cuvillier.de

Members of the Thesis Committee

Prof. Dr. Dirk Görlich   Max Planck Institute for Biophysical Chemistry
(Supervisor and referee)   Department of Cellular Logistics
                                           Göttingen, Germany

Prof. Dr. Peter Rehling   Georg August University Göttingen
(Co-referee)   Department of Biochemistry II
                                           Göttingen, Germany

Prof. Dr. Helmut Grubmüller   Max Planck Institute for Biophysical Chemistry
                                           Department of Theoretical and Computer-based
                                           Biophysics
                                           Göttingen, Germany

Additional Members of the Examination Board

Prof. Dr. Blanche Schwappach   Georg August University Göttingen
                                          Department of Biochemistry I
                                          Göttingen, Germany

Prof. Dr. Silvio Rizzoli   European Neuroscience Institute
                                          Research Group STED Microscopy of Synaptic
                                          Function
                                          Göttingen, Germany

Prof. Dr. Detlef Doenecke   Georg August University Göttingen
                                          Department of Biochemistry I
                                          Göttingen, Germany

Affidavit

I hereby declare that my dissertation, entitled '**Evolutionary conserved features of FG repeats that allow the formation of hydrogel-based permeability barriers with NPC-like properties**', has been written independently and with no other aids or sources than quoted.

Moreover, I attest that this dissertation has not been submitted elsewhere for the pursuit of any academic award or qualification.

Hermann Broder Schmidt
Göttingen, September 2012

To thee, who have taught me

Table of Contents

List of Publications

Ader, C., Frey, S, Maas, W., **Schmidt, H.B.**, Görlich, D. and Baldus, M. (2010).
Amyloid-like interactions within nucleoporin FG hydrogels.
Proceedings of the National Academy of Sciences 107, 6281–6285.

List of Figures

List of Tables

List of Abbreviations

A_{280}	Absorbance at λ = 280nm.
AA	Amino acid
ADP	Adenosine 5'-diphosphate
ATP	Adenosine 5'-triphosphate
BSA	Bovine serum albumin
CAS	Cellular apoptosis susceptibility (also referred to in the literature as Exportin-2).
C-term.	Carboxy-terminal
DNA	Deoxy-ribonucleic acid
DTT	Dithiothreitol
E. coli	*Escherichia coli*
EDTA	Ethylenediaminetetraacetic acid
Exp	Exportin
FG repeat	Phenylalanine (**F**)-glycine (**G**) repeat
FOA	5'-Fluoroorotic Acid
GDP	Guanosine 5'-diphosphate
GFP	Green fluorescent protein
GHCl	Guanidine hydrochloride
GTP	Guanosine 5'-triphosphate
GTPase	GTP hydrolase
HPLC	High performance liquid chromatography
IBB	Impβ-binding domain of Impα
IMA	Importin-alpha
Imp	Importin
IPTG	Isopropyl-β-D-thiogalactopyranoside
kDa	kilo Dalton
LCEA	Last common eukaryotic ancestor
MBP	Maltose-binding protein
mCherry	monomeric Cherry
NPC	Nuclear pore complex
NTR	Nuclear transport receptor
Nup	Nucleoporin
o/n	overnight
OGT	O-β-N-acetylglucosaminyltransferase
PBS	Phosphate-buffered saline
PEG	Polyethylene glycol
PMSF	Phenylmethylsulfonylfluoride
PrD	Prion-determining domain
Ran	Ras-related nuclear antigen
RanBP	Ran-binding protein
RanGAP	RanGTPase-activating protein

RanGEF	Ran guanidine nucleotide exchange factor
Rcc1	Regulator of chromosome condensation 1
RNA	Ribonucleic acid
RNP	Ribonucleoprotein
RT	room temperature
scRNA	scan ribonucleic acid
SDS-PAGE	Sodium dodecyl sulfate-polyacrylamide gel electrophoresis
TB	Terrific broth medium
tCherry	tetrameric Cherry
TEV	Tobacco etch virus
TRIS	Tris(hydroxymethyl)aminomethane, 2-amino-2-hydroxymethyl-1,3-propandiol
v/v	volume per volume
WT	Wilde type
w/v	weight per volume
YT / 2YT	Yeast extract tryptone medium
Z (in ZZ)	IgG-binding domain of the *Staphylococcal* protein A

Standard single-letter amino acid codes and the Interantional System of units (SI) were used.

Abstract

The permeability barrier of nuclear pore complexes is formed by non-globular phenylalanine-glycine (FG) repeat domains and controls the exchange of macromolecules between nucleus and cytoplasm. It allows for two modes of passage: passive diffusion, which is efficient only for molecules below a size limit of ~30kDa, and facilitated transport of larger cargo molecules chaperoned by nuclear transport receptors (NTRs).

According to the selective phase model, FG repeats interact to form a three-dimensional, sieve-like meshwork that constitutes the NPC permeability barrier. Accumulating evidence indeed points to a 'cohesive' nature of FG repeats, remarkably enabling them to form a 'selective phase' *in vitro*: macroscopic hydrogels that reproduce the fundamental aspects of nucleocytoplasmic transport. If such a phase however was the essential feature of the NPC permeability barrier, the cohesiveness of FG repeats and their ability to form selective phases should be evolutionary conserved in all eukaryotes. To test this hypothesis, FG repeats were systematically identified in every eukaryotic clade, and the properties of selected FG repeats from representative model organisms explored in detail.

The analysis of the complete database revealed that although FG repeats are generally rich in glycine, serine, threonine and phenylalanine residues, their overall hydrophobicity and content of asparagine (N) and glutamine (Q) side chains negatively correlated. Interestingly, the hydrophobic FG repeats from vertebrates and plants formed hydrogels that were excellent passive barriers towards inert molecules, but hardly allowed the enrichment and intra-gel diffusion of large nuclear transport receptor (NTR)•cargo complexes. In contrast, the NQ-rich FG repeats e.g. from yeast constituted hydrogels that showed an 'incorrect' affinity for inert molecules, which however was by far less strong than their interaction with NTR•cargo complexes. These findings suggest that there are two extreme solutions of making FG repeats cohesive, which each come with their own intrinsic side effects.

Strikingly, the data presented in this study indicates that the unicellular eukaryote *Tetrahymena thermophila*, which contains two morphologically and functionally distinct nuclei in a common cytoplasm, might make use of these different cohesiveness modes to operate permeability barriers with orthogonal specificities.

Finally, the previously suggested relationship between NQ-rich prions and FG repeats was further solidified in this study by converting the amyloid-forming Sup35 prion-determining domain into an FG repeat-like domain capable of forming selective hydrogels.

1. Introduction

One of the most fundamental principles of life is the ability to store, process and transmit the information required to synthesize the macromolecular building blocks, perform the biochemical reactions and coordinate the cellular processes that constitute the cell. Although all cells store this information in form of DNA, transcribe it into various forms of RNA (in particular rRNAs, mRNAs, tRNAs or regulatory RNAs) and eventually translate some of it into proteins (with mainly architectural or enzymatic functions), the spatiotemporal arrangement of this flow of information shows remarkable differences in pro- and eukaryotes. This is due to the fact that eukaryotes contain, amongst many other sub-cellular compartments, a cell nucleus that hosts the genome.

The nucleus is delimited by the nuclear envelope (NE), a double membrane that is continuous with the endoplasmic reticulum (ER). Consequently, the key processes of transcription (in the nucleus) and translation (by cytoplasmic ribosomes) are separated by a physical barrier in eukaryotic cells, no longer allowing the immediate co-transcriptional translation of an emerging transcript into protein. Rather, an additional exchange step between nucleus and cytoplasm is required to circumvent an interruption of the continuous flow of information in eukaryotes. Amongst the even more profound consequences of arranging the genetic information into a nucleus is the need for a fundamentally new mechanism to segregate the duplicated DNA after replication (i.e. mitosis), as the DNA is no longer attached to the cytoplasmic membrane and thus cannot simply be segregated by surface membrane motors anymore (Cavalier-Smith, 2010b).

Yet, the cellular sub-organization into a nuclear and cytoplasmic compartment clearly offered manifold opportunities in regard to genome maintenance and regulation of gene expression. Firstly, the confinement of the genetic information into an enveloped and mechanically supported compartment greatly contributes to the stability of the genome, hence allowing eukaryotes to manage ~1000-fold larger genomes than prokaryotes. Indeed, not only the size, but secondly also the structure of the genome dramatically changed from (predominantly) bacterial operons containing multiple open reading frames that are transcribed into polycistronic mRNAs to the typical intron/exon structure of modern eukaryotic genes, which are first transcribed into precursor mRNAs and eventually spliced into mature monocistronic mRNAs (Alberts et al., 2002). Importantly, the spatiotemporal separation of transcription (and concomitantly RNA processing) from translation was an inevitable prerequisite for (alternative) mRNA splicing to evolve and manifest itself as a mechanism to greatly enhance the coding potential of eukaryotic genomes, since the translation of un- or incompletely spliced transcripts, which could give rise to non-functional proteins or proteins with dominant-negative effects, is thereby elegantly prevented (as e.g. also discussed in Görlich and Kutay, 1999). In this regard, it is also noteworthy that individual exons in general encode for independent protein domains and that recombination of previously unlinked exons via exon shuffling thus greatly favored the evolution of multi-domain proteins with diverse functions and specificities (Gilbert, 1978). Thirdly, the possibility to control the access of regulatory molecules such as transcription factors to the genetic information allows to fine-tune gene expression, e.g. in response to specific intra- or extracellular signals. Finally, a general benefit of compartmentalization is the local concentration of specific factors, thus increasing the efficiency of the catalytic reactions conducted in a given cellular organelle. Taken together, these new possibilities played a pivotal role in the evolution of eukaryotes into complex multicellular organisms.

1.1 The Fundamental Aspects of Nucleocytoplasmic Exchange

Evidently, the information stored and processed in the nucleus has to reach the cytoplasm to be read and interpreted. Therefore, mature RNA species (especially protein-coding mRNAs) and ribonucleoproteins (RNPs; e.g. fully assembled ribosomal subunits) have to be exported to the cytoplasm, whereas all nuclear proteins such as histones, transcription factors or the components of the DNA replication machinery need to be imported into the nucleus. Indeed, nucleocytoplasmic transport is characterized by an immense bidirectional mass flux (estimated to amount to $\approx$10-20 MDa$\cdot$NPC$^{-1}\cdot$s^{-1}; Ribbeck and Görlich, 2001) and demands considerable cellular resources. Taken together, the complete nuclear transport machinery comprises (i) nuclear pore complexes (NPCs), which are large proteinaceous gateways allowing for nucleocytoplasmic exchange, (ii) nuclear transport receptors (NTRs) dedicated to chaperone selected molecules into or out of the nucleus and (iii) the RanGTPase system, which feeds in metabolic energy, thereby bestowing directionality to the transport processes.

All exchange of macromolecules between nucleus and cytoplasm proceeds solely through **NPCs** (Feldherr et al., 1984), which perforate the nuclear envelope (Watson, 1954). NPCs are large, structurally conserved assemblies built from multiple copies of approximately 30 different proteins (Rout et al., 2000; Cronshaw et al., 2002) called nucleoporins (Nups). They comprise a rigid core scaffold with eightfold rotational symmetry (Wischnitzer, 1958; Gall, 1967) and a central channel equipped with a permeability barrier.

Importantly, the permeability barrier allows two modes of passage through the nuclear pore: (i) passive diffusion, which is efficient only for small molecules below a size limit of 40 kDa (Harding and Feldherr, 1958; Bonner, 1975; Mohr et al., 2009), and (ii) facilitated transport of even larger cargo molecules (reviewed e.g. in Görlich and Kutay, 1999; Fried and Kutay, 2003). In order to handle the vast nucleocytoplasmic mass fluxes, each NPC accommodates for ~1,000 facilitated translocation events per second, turning over the equivalent of its own mass (~125kDa in higher eukaryotes) almost every second (Ribbeck and Görlich, 2001).

Facilitated transport relies on **NTRs** (Moore and Blobel, 1992; Görlich et al., 1994; Imamoto et al., 1995; Görlich et al., 1995a; 1995b; Pollard et al., 1996; Görlich, 1997; Fornerod et al., 1997a), which shuttle between the nucleus and cytoplasm. NTRs bind to specific import (e.g. the classical nuclear localization signal (NLS) of the simian virus 40 large T-antigen, Kalderon et al., 1984) or export (e.g. the classical leucine-rich nuclear export signals, see also Güttler et al., 2010) signals on the cargo molecules and mediate their translocation through the nuclear pore.

Most NTRs belong to the Impβ superfamily and hence share a number of features such as their relatively large size (typically 90-150kDa), their overall negative charge and their ability to bind the small guanine nucleotide-binding protein Ran (Görlich et al., 1997; Fornerod et al., 1997a). Indeed, Impβ-like NTRs are even build up out of the same basic structural units (reviewed e.g. in Cook et al., 2007; Güttler and Görlich, 2011), so called HEAT repeats (for *h*untingtin, *e*longation factor 3, the PR65/*A* subunit of protein phosphatase 2*A* and the lipid kinase *T*OR, which were the first proteins identified to posses this structural motif). One HEAT repeat unit is comprised of two antiparallel alpha helices connected by a short linker. Multiple HEAT repeats are then stacked in a tandemly arranged fashion into superhelical solenoids, giving Impβ-like NTRs their characteristic shape.

A complete translocation round for Impβ-type NTRs (in the following referred to as NTRs for simplicity) comprehends six defined steps (see also Figure 1.1): (i) recognition of cargoes by cognate NTRs, (ii) docking of the NTR•cargo complexes to the NPC and penetration into the permeability barrier, (iii) the actual translocation through the permeability barrier, (iv) exit from the permeability barrier, (v) NTR•cargo complex disassembly and (vi) retrieval of the NTRs to the starting compartment (strictly seen including another docking, translocation and exiting step). According to the directionality of the transport routes they participate in, NTRs can be classified as either Importins (Imps) or Exportins (Exps), whereas only few are known to function as both an Importin or Exportin (see Table 1.1).

Table 1.1. *Nuclear transport receptors of the Impβ-family and their cargoes.*[1]

– Importins –		
NTR	Adaptor or co-receptor	Selected cargoes
		Ribosomal proteins
		HIV Rev, HIV Tat
		Histones
Importin β	Importinα	Classical NLS-containing cargoes
(Impβ-1)	Snurportin 1	m_3G-capped U-snRNPs
	XRIPα	Replication protein A
	Importin 7	Histone H1

NTR	Adaptor or co-receptor	Selected cargoes
Transportin 1+2 (Trn, Impβ-2)		hnRNP proteins (M9-NLS) Ribosomal proteins
Transportin 1+2 (Trn, Impβ-2)		TAP Histones c-FOS SRP19
Transportin SR 1+2 (TrnSR, Trn3)		SR proteins tRNA
Importin 4		Ribosomal proteins Histones
Importin 5		Ribosomal proteins Histones
Importin 7	Importin β	Ribosomal proteins Histone H1 ERK2, SMAD3, MEK1
Importin 8		SRP19 Argonaute proteins
Importin 9		Histones Ribosomal proteins
Importin 11		UbcM2 Ribosomal protein L12

– *Exportins* –		
NTR	**Adaptor or co-receptor**	**Selected cargoes**
Crm1 (Exportin 1)	HIV Rev PHAX	Leu-rich NES cargoes RRE-containing RNAs m^7G-capped U snRNAs Snurportin 1
CAS (Exportin 2)		Importin αs
Exp-t		tRNA
Exportin 5	aa-tRNA dsRNA	eEF1A dsRNA-binding proteins pre-miRNAs
Exportin 6		Actin•profilin complexes
Exportin 7		p50 RhoGAP, 14-3-3σ

– *Bi-directionally operating NTRs* –		
NTR	**Adaptor or co-receptor**	**Selected cargoes**
Importin 13		UBC9, MGN/Y14 (import) eIF1A (export)
Exportin 4		eIF5A (export) SMAD3 (export) Sox2, SRY (import)

[1] Compiled based on (Görlich and Kutay, 1999; Fried and Kutay, 2003; Güttler and Görlich, 2011).

Notably, Impβ does not directly bind to many of its import cargoes, hence undergoing a slightly more complicated translocation cycle including an additional component (for a detailed review, see Görlich and Kutay, 1999). The latter is the import adaptor Impα, which recognizes the classical NLSs of a broad range of cargo molecules in complex with Impβ (Görlich et al., 1994; 1995a; 1995b). A separate high-affinity signal sequence on Impα, the importin β-binding domain (IBB), mediates the interaction with Impβ (Görlich et al., 1996a).

Imp-α proteins are build from armadrillo (ARM) motifs (Görlich and Kutay, 1999), which are structurally related to the HEAT repeats of *bona fide* NTRs of the Imp-β subfamily (Andrade et al., 2001). However, even though Imp-α is likely to have a higher 'solubility' in the NPC permeability barrier than truly inert proteins (see Section 1.3), it is nevertheless a poor mediator of facilitated translocation independent of Imp-β (Görlich et al., 1995a; 1995b).

An important property of NTRs is their regulation by **Ran**. Like related small GTP-binding proteins, Ran functions as a molecular switch residing either in an ON or OFF state, depending on whether GTP or GDP is bound, respectively. The change of states is accompanied by large conformational changes in Ran (see also Figure 2.4 and Scheffzek et al., 1995; Vetter et al., 1999; Partridge and Schwartz, 2009), which can be transmitted to interacting NTRs.

As the intrinsic nucleotide hydrolysis capability of Ran is very low, the molecule cannot switch states on its own, but rather requires the help of the GTPase-activating protein RanGAP (in conjunction with the Ran binding proteins RanBP1 or RanBP2/Nup358) to hydrolyze the bound GTP (Bischoff et al., 1994; Bischoff and Görlich, 1997) and the guanine nucleotide exchange factor Rcc1 to exchange GDP for GTP (Bischoff and Ponstingl, 1991a).

Due to the restriction of RanGAP, RanBP1 and RanBP2/Nup358 to the cytoplasm or cytoplasmic face of the NPC and the chromatin-association of Rcc1, a steep RanGTP gradient is establish across the nuclear envelope, with high nuclear and low cytoplasmic concentrations (Görlich et al., 1996b; Izaurralde et al., 1997). Thus, Ran can convey information about the identity of the compartment to the shuttling NTRs, which *per se* can traverse the nuclear pore (even in a cargo-bound state) in either direction in an energy-independent manner (Kose et al., 1997; Schwoebel et al., 1998; Nachury and Weis, 1999; Ribbeck et al., 1999).

In molecular terms, Importins and Exportins respond diametrically opposite to binding of RanGTP (Figure 1.1). For Importins, which form dimeric import complexes with their cargoes in the cytoplasm, RanGTP binding causes the dissociation of the import complex and release of the cargo in the nucleus (Rexach and Blobel, 1995; Görlich et al., 1996b). Exportins however can only bind their cargoes with affinity when engaging in a trimeric export complex with RanGTP (Fornerod et al., 1997a; Kutay et al., 1997a). RanGAP- and RanBP1/2-aided nucleotide hydrolysis in the cytoplasm in turn leads to export complex disassembly and cargo release. Thus, the chemical potential of the RanGTP gradient is the driving force of directional nucleocytoplasmic transport

(Izaurralde et al., 1997; Görlich et al., 2003), moreover allowing the accumulation of transport substrates against gradients of chemical activity.

Note however that active nuclear transport would soon deteriorate the RanGTP gradient over time if the RanGDP that accumulates in the cytoplasm due to export complex disassembly and retrieval of Importins were not re-imported into the nucleus and converted into RanGTP again. This duty is fulfilled by NTF2 (Ribbeck et al., 1998; Smith et al., 1998), a dedicated import receptor for RanGDP (Stewart et al., 1998) importantly not belonging to the Impβ superfamily (Bullock et al., 1996). Hence in this case, cargo release can be independent of RanGTP binding, an essential requirement to abstain from futile cycles of Ran shuttling. Instead, RanGDP release is believed to rely on the concerted action of Rcc1 and Impβ-like NTRs (Ribbeck et al., 1998; Smith et al., 1998).

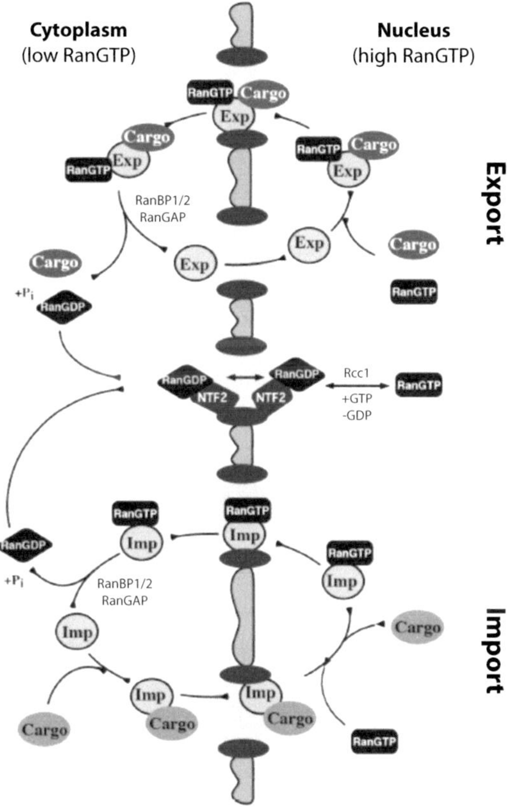

Figure 1.1. Schematic overview of the nuclear export and import cycles and their regulation by the RanGTPase system. Adapted from (Görlich and Kutay, 1999). 'Exp' denotes Exportin and 'Imp' stands for Importin. See main text for further details.

In summary, the NPC permeability barrier efficiently restricts the diffusion of small proteins (i.e. below >5nm in diameter) into and out of the nucleus, yet paradoxically, it requires the help of large NTRs even for similar sized cargoes to surmount this size barrier in a reasonable time. In turn, how such a barrier operates and what the physical nature of the NPC permeability barrier is, remain the major unsolved and highly debated questions in the field of nucleocytolasmic transport, and essentially motivated this study. In the following, the molecular architecture of the NPC (Section 1.2) and current models of NPC permeability barrier function (Section 1.3) will be introduced in more detail to summarize the collective proceedings of the field and to deduce the specific aims of this work.

1.2 The Nuclear Pore Complex

The concepts of nucleocytoplasmic transport differ in many aspects from other intracellular transport pathways, such as protein import into the ER, mitochondria, chloroplasts or peroxisomes (to name a few). Maybe most importantly, small molecules and metabolites, as well as ions, can freely exchange between nucleus and cytoplasm, whereas great care is taken so that the specific, fine-tuned biochemical milieu of other compartments is not disturbed due to material exchange with the cytoplasm. This is because molecules do not have to be transported across a lipid bilayer to enter or exit the nucleus. Rather, the outer and inner membranes of the NE are fused at the sites where NPCs reside, leaving an aqueous passage that is however guarded by the NPC permeability barrier. Hence, NPCs function both as 'grommets' that reinforce the pores in the NE and selective gates for bidirectional nucleocytoplasmic exchange.

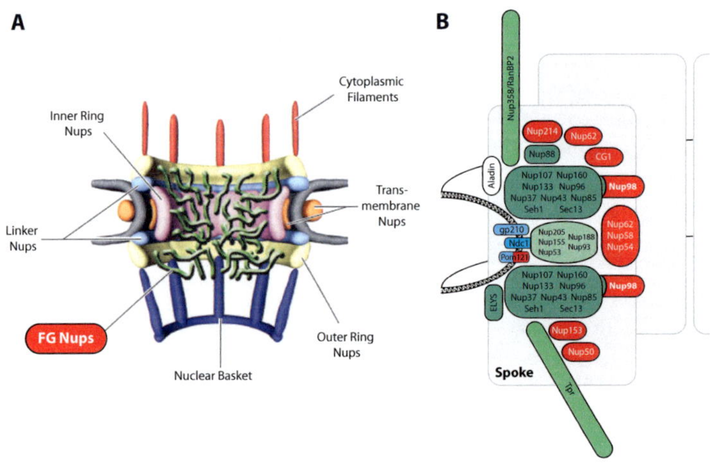

Figure 1.2. The architecture of the nuclear pore complex (NPC). (A) Schematic drawing of the NPC (adapted from Grossman et al., 2012), illustrating how the outer ring, linker, inner ring and transmembrane Nups form the core NPC scaffold. The permeability barrier is constituted by the FG Nups filling the central channel of the NPC. Extending into the cytoplasm are the cytoplasmic filaments. A basket-like structure is found at the nuclear face of the NPC. (B) The basic symmetry unit of the NPC is a 'spoke'. Shown here are the Nup subcomplex that comprise a single spoke. Please note that eight spokes align in the equatorial plane of the nuclear envelope to form the multiple coaxial rings of the core scaffold. See main text for details, especially on the Nup subcomplexes.

1.2.1 The NPC is a highly modular macromolecular assembly.

Different parts of the highly modular NPC structure serve these two purposes (Figure 1.2). The core scaffold of the NPC, which comprises two outer and two inner coaxial rings, forms the grommet-like structure (reviewed e.g. in Grossman et al., 2012). It is anchored to the NE via integral membrane Nups that interact with components of the inner ring (Onischenko et al., 2009). Extending from the outer rings into the cytoplasm and nucleus are eight cytoplasmic filaments and a nuclear basket-like structure, respectively (Jarnik and Aebi, 1991; Goldberg and Allen, 1992). The inner walls of the coaxial rings are lined with Nups containing so-called phenylalanine-glycine (FG) repeat domains, which are thought to form the permeability barrier (see below for details).

Electron microscopy studies of *Xenopus* (Unwin and Milligan, 1982; Hinshaw et al., 1992; Akey and Radermacher, 1993; Akey, 1995; Stoffler et al., 2003; Frenkiel-Krispin et al., 2010), *Dictyostelium* (Beck et al., 2004; 2007), yeast (Yang et al., 1998; Kiseleva et al., 2004; Alber et al., 2007) and human (Maimon et al., 2012) NPCs showed that their general architecture is well conserved. The reported size estimates of the NPC range from ~65MDa to ~125MDa, depending on both the species analyzed and the methods used (Reichelt et al., 1990; Rout and Blobel, 1993).

The basic symmetry unit of the NPC is referred to as a spoke, which is constituted by distinct biochemically defined sub-complexes of either (i) transmembrane, (ii) core scaffold or (iii) FG repeat domain Nups (Figure 1.2). Eight spokes align in an equatorial plane to assemble into the multiple coaxial rings of the core scaffold, explaining the characteristic eightfold rotational symmetry already observed in the first electron micrographs of NPCs (Gall, 1967). Moreover, this organization also requires that each Nup is present in single or multiple copies of eight. Taking into consideration that NPCs are comprised of ~30 distinct Nups (Rout et al., 2000; Cronshaw et al., 2002), at least ~240 protein molecules are required to build the complete structure. The exact copy numbers of the different Nups are still largely under debate, but it is assumed that NPCs are composed of as many as ~700 individual molecules (Rout et al., 2000). Despite this large number of contributing parts, the symmetry constraints, the organization of most Nups into sub-complexes and the findings that Nups are only made up of limited domain topologies greatly reduce the overall (structural) complexity of the NPC.

1.2.2 Defined sub-complexes are the major building blocks of higher order NPC structure.

In context of the NPC, individual sub-complexes are defined based on the ability of a subset of Nups to stably interact with each other, even when the nuclear envelope breaks down during open mitosis. Indeed, NPCs reassemble from the fragmented sub-complexes at the end of mitosis (reviewed in Dangelo and Hetzer, 2008), highlighting their importance as the major building blocks of higher order structural elements (i.e. single spokes). Notably, based on mainly biochemical analysis of homologous components, the same sub-complexes can also be identified in eukaryotes undergoing closed mitosis (e.g. yeast). Historically, Nups are named after their molecular weight, which might however vary for homologous proteins from distantly related organisms. In the following, the mammalian nomenclature is used when referring to specific Nups, unless explicitly stated otherwise. For comparison, Table 1.2 gives an overview of the known Nups identified in various Eukaryotes. Notably, recent progress in genome sequencing and improvement of algorithms for homology detection suggest that the different sub-complexes are evolutionary well conserved (Mans et al., 2004; Bapteste et al., 2005; Neumann et al., 2010).

Table 1.2. *Vertebrate nucleoporins and their (known) homologs in selected species.*

Localization	Vertebrates[1]	Yeast[1] (S. cerevisae)	Tetrahymena thermophila[2]	Trypanosoma brucei[3]
Cytoplasmic fibrils	Nup88	Nup82		
	Nup214	Nup159	Nup192?	
	Nup358/RanBP2	-		Nup308?
Transmembrane	Ndc1	Ndc1		
	Pom121	-		
	gp210	Pom152	pp210	
Outer Ring	Nup85	Nup85		
	Nup96	Nup145C	Nup96	Nup158
	Nup107	Nup84		Nup82/Nup89
	Nup133	Nup133		Nup109/Nup132
	Nup160	Nup120		Nup109/Nup132
	Sec13	Sec13	Sec13	Sec13
	Seh1	Seh1	Seh1	
	Nup37	-		
	Nup43	-		
	ELYS	-		
	Aladin	-		Nup48
Inner Ring	Nup53	Nup53		
	Nup93	Nic96	Nup93	Nup96

Inner Ring	Nup155	Nup157/Nup170	Nup155	Nup144
	Nup188	Nup188		Nup181/Nup225
	Nup205	Nup192	Nup199?	Nup181/Nup225
Central channel	Nup54	Nup57	Nup54	Nup53a/Nup53b?
	Nup58	Nup49		Nup58?
	Nup62	Nsp1		Nup62?
	Nup98	Nup100/Nup116/ Nup145N	MacNup98A/B MicNup98A/B	Nup158
	CG1	Nup42		
Nuclear Basket	Tpr	Mlp1/Mlp2		Nup92/Nup110
	Nup50	Nup2	Nup50	
	Nup153	Nup1		

[1] For a comparison of vertebrate and yeast Nups, see e.g. (Grossman et al., 2012).
[2] See (Malone et al., 2008; Iwamoto et al., 2009) and Table 1.3.
[3] See (Degrasse et al., 2009).

The major constituent of the outer ring is the heptameric **Y complex**, named after its characteristic Y-shape observed in electron micrographs (Siniossoglou et al., 2000; Lutzmann et al., 2002). It comprises the major components Nup85, Nup96, Nup107, Nup133, Nup160, Sec13 and Seh1, plus the additional proteins Nup37, Nup43 and ELYS in many eukaryotes (excluding for example the yeast *Saccharomyces cerevisiae*, which nevertheless is a well established model organism in the field). Associated with the Y complex is the FG repeat domain-containing nucleoporin (FG Nup) **Nup98**, which interacts directly with Nup96. Importantly, Nup98 is considered to be the major constituent of the NPC permeability barrier (see also Section 1.3 and (Laurell et al., 2011; Hülsmann et al., 2012). Interestingly, Nup98 is post-translationally O-glycosylated in vertebrates such as *Xenopus laevis* and *Homo sapiens* (Powers et al., 1995; Radu et al., 1995).

The inner ring is formed by the **Nup93-Nup205 sub-complex**, which mainly consists of Nup93, Nup188 and Nup205 (Grandi et al., 1997; Miller et al., 2000), but in some models also contains Nup53 and Nup155 (Hawryluk-Gara et al., 2005), through which interaction the entire sub-complex is linked to the transmembrane Nups Ndc1 (Mansfeld et al., 2006; Stavru et al., 2006), Pom121 (Hallberg et al., 1993) and gp210 (Gerace et al., 1982). The Nup93-Nup205 sub-complex is not as defined as the well-studied Y complex, mainly because the interactions between the two constituting parts are well documented, but not as stable as the interactions within the parts. Thus, in *S. cerevisiae* for example, the Nup35 and Nup155 homologs are sometimes considered to form an independent sub-complex. Attached to the Nup93-Nup205 sub-complex is the **Nup62 sub-complex** (Finlay et al., 1991), which is hence located to the center of the NPC. It comprises the FG nucleoporins Nup54, Nup58 and Nup62. Like Nup98, the latter is O-glycosylated in vertebrates (Davis and Blobel, 1987).

The cytoplasmic filaments are comprised of the **Nup214-Nup88 sub-complex** (Kraemer et al., 1994; Bastos et al., 1997; Fornerod et al., 1997b), and also contain

Nup358/RanBP2 in metazoa (Bernad et al., 2004). Thus, in higher eukaryotes, export complex disassembly is directly localized to the cytoplasmic face of the NPC (see above). Notably, it was shown that the Nup214-Nup88 sub-complex provides also additional binding sites for Nup62 (Belgareh et al., 1998). The key component of the nuclear basket is **Tpr** (Krull et al., 2004). It is required to 'clear' the vicinity of NPCs of heterochromatin (Krull et al., 2010). Associated with the nucleoplasmic face of the NPC are the FG Nups Nup50 and Nup153 (Sukegawa and Blobel, 1993).

1.2.3 Scaffold nucleoporins resemble membrane coat proteins.

This seemingly detailed understanding of the NPC components and overall architecture oftentimes overshadows the fact that high-resolution structures are only available for few Nups. Indeed, understanding of the NPC in atomic detail is a major focus of recent efforts in the field (Brohawn et al., 2009). From the solved structures, in combination with biochemical characterization and structural predictions, it has however become evident that Nups can be grouped into three main fold classes. Accounting for most of the NPC mass (~75% in yeast) are α-solenoids, β-propellers and tandem combinations of both folds. The second category is comprised of Nups containing coiled-coil domains. Lastly, approximately one-third of all Nups contain FG repeat domains.

Structurally particularly well studied is the Y complex. Like all core scaffold Nups, also the members of this complex are mainly comprised of α-solenoids and/or β-propellers (reviewed in Brohawn et al., 2009). Especially the latter are often-encountered folds, found in many proteins with diverse functions. However, the combination of N-terminal β-propeller and C-terminal α-solenoid domains is very rare and has so far only been found in a subset of membrane coat proteins (with Sec31 being the most prominent example) and some Nups such as Nup133, Nup155 and Nup160 (Devos et al., 2004; Leksa et al., 2009; Whittle and Schwartz, 2009; Santarella-Mellwig et al., 2010). Whereas in these examples, both modules are combined in a single polypeptide chain, there are also examples within the NPC scaffold where an otherwise solely α-solenoid Nup (e.g. Nup85 or Nup96) contributes the missing blade to the β-propeller fold of its interaction partner (Seh1 or Sec13, respectively), thereby yielding the same overall fold (Brohawn et al., 2008; Brohawn and Schwartz, 2009). Based on these similarities, the protocoatomer hypothesis was proposed, suggesting that certain NPC components and clathrin, COPI and COPII vesicle coats share a common evolutionary origin (Devos et al., 2004; Degrasse et al., 2009; Field et al., 2011).

Moreover, also coiled-coil interactions play an important role in sub-complex formation and anchoring. For example, the three members of the Nup62 sub-complex associate via such interactions, with an additional coil being contributed by Nup93, thereby anchoring the entire sub-complex to the inner ring of the NPC core (Buss and Stewart, 1995; Grandi et al., 1995; Hu et al., 1996; Solmaz et al., 2011). Coiled-coils are also the dominating structural motif of the nuclear basket (Byrd et al., 1994; Krull et al., 2004).

1.2.4 The NPC permeability barrier is formed by FG repeat domains.

Apart from the transmembrane and core scaffold Nups, NPCs are built form a third category of crucial components, namely FG Nups. These contain characteristic FG repeat domains (or short FG repeats), which have been first identified in the *Saccharomyces cerevisiae* nucleoporins Nsp1 (Nehrbass et al., 1990) and Nup1 (Davis and Fink, 1990). In general, FG repeat domains are non-globular (Denning et al., 2003) and comprise up to 50 individual FG repeat units, in which a hydrophobic FG, FxFG or GLFG motif is embedded into a more hydrophilic spacer sequence. As described above, FG repeats are anchored to the NPC core scaffold via terminal structured domains (see also Section 2.1.1).

FG repeats serve as docking sites for NTRs (e.g. Iovine et al., 1995; Paschal and Gerace, 1995; Radu et al., 1995; Rexach and Blobel, 1995; Fornerod et al., 1997b), with structural information on their interaction available (Bayliss et al., 2000; 2002a; 2002b). Importantly, the binding of NTRs to the hydrophobic patches of the FG repeats is essential for facilitated NPC passage (Bayliss et al., 1999).

It is generally accepted that FG repeats are essential for viability (Strawn et al., 2004) and critical for the permeability barrier (Frey and Görlich, 2007; Patel et al., 2007; Hülsmann et al., 2012).

The vertebrate NPC encompasses 10 FG repeat domain-containing Nups, namely Nup54, Nup58 and Nup62 (i.e. the Nup62 subcomplex), Nup98 and CG1 (all five directly localized to the central channel), as well as Nup50, Nup153 (on the nucleoplasmic face of the NPC) and Nup214, Nup358 (as part of the cytoplasmic fibrils). Additionally, the integral membrane Nup Pom121 also contains FG repeats. Notably, comprehensive proteomic studies revealed that also in the yeast and Trypanosome NPC, approximately one-third of all Nups contain FG repeats (Rout et al., 2000; Degrasse et al., 2009). Altogether, it has been estimated that more than 1000 FG repeats are present in each NPC, resulting in an effective concentration of ~50mM in the central channel (Bayliss et al., 1999). Thus, FG repeats are a highly abundant feature of the nuclear pore.

Taken together, compelling evidence suggests that FG repeats are the physical constituents of the NPC permeability barrier. In contrast to the scaffold Nups however, FG repeats can neither be seen in electron microscopy nor studied by high resolution structural methods due to the their lack of a defined folding and concomitant inherent flexibility. Hence, there is also no direct structural information on the permeability barrier available to date, let alone on the interactions between the permeability barrier and a translocating species. Indeed, merely small fragments of FG repeats spanning isolated

NTR binding sites have been co-crystalized with NTRs (Bayliss et al., 2000; 2002a; 2002b), by far not representing the complex interactions between NTRs (which contain multiple bindings sites for FG motifs) and FG repeats (which contain many FG motifs) during facilitated translocation (Bednenko et al., 2003; Morrison et al., 2003; Isgro and Schulten, 2005). In summary, the key question is how the binding of NTRs to FG repeats is coupled to a selective barrier passage.

1.3 Models for the NPC Permeability Barrier

Mechanistically, it is not trivial to explain how FG repeats can promote facilitated translocation by engaging in multivalent interactions with NTRs and at the same time hinder passive diffusion of inert species. Importantly, the permeability barrier of NPCs is first and foremost a size-dependent kinetic barrier, meaning that increasingly larger molecules take progressively longer to pass through nuclear pores (Figure 1.3). NTRs in turn can accelerate the passage times of their cargoes by a factor of ~1000 (Ribbeck and Görlich, 2001). Intuitively however, one would predict that a mere binding of NTRs to FG repeats should lead to the retention of any translocating species and consequently cause a delay of its passage through the NPC, rather than facilitating it.

Over the past ten years, several models have been postulated and intermittently refined to resolve this paradox and explain the nature and function of the NPC permeability barrier. These models conceptually differ in their predictions of both the physical arrangement of FG repeats within the NPC (non-interacting, extended bristles vs. entangled polymer brushes vs. reversibly cross-linked meshworks) and the actual selectivity/translocation mechanism (entropic gate vs. reversible collapse vs. reduction of dimensionality vs. molecular sieving).

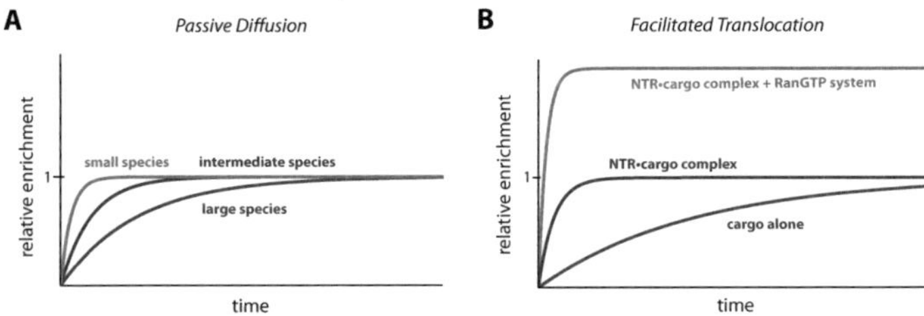

Figure 1.3. The NPC permeability barrier operates as a size-dependent kinetic barrier. (A) Consider three differently sized cytoplasmic proteins with a step concentration gradient across the nuclear envelope, which all do not possess interaction partners in the nucleus and hence have a final partitioning coefficient of 1.0 (resembling equilibration between the two compartments). The time it takes for these molecules to equilibrate depends on their size (see e.g. Mohr et al., 2009): the larger a given species is, the smaller its NPC passage rate will be. (B) The same is true for large nuclear proteins as well. Remarkably however, the NPC passage rate of nuclear proteins can be drastically increased when chaperoned by NTRs, despite the further increase in total size of the translocating species (see e.g. Ribbeck and Görlich, 2002). Note that with the help of the RanGTP system, nuclear proteins can even be enriched against concentration gradients, reaching final partitioning coefficients well above 1.0.

1.3.1 Virtual gating and the Brownian affinity gating model

The virtual gating hypothesis is essentially a thermodynamic argument for the existence of an entropic barrier guarding the nuclear pores. It is based on the fact that for any molecule freely diffusing in the cytoplasm or nucleoplasm, entering the confined space of the NPC is accompanied with high entropic costs, which are further exacerbated by the presence of natively unfolded FG repeats acting as 'entropic bristles' (Rout et al., 2003). Especially molecules exceeding the known size limit for efficient passive diffusion are thought to be so severely effected, that they are 'virtually' denied access. An exception are NTRs, which engage in multiple low-affinity interactions with FG repeats, thereby allegedly releasing sufficient binding enthalpy to pay for the entropic price associated with translocating through the NPC (Rout et al., 2003). Note that the off-rates for individual binary NTR-FG repeat associations are high enough to prevent significant retention of a translocating species (Rout et al., 2000; Ribbeck and Görlich, 2001; Rout et al., 2003; Tetenbaum-Novatt et al., 2012).

The idea of virtual gating was first conceived in conjunction with the Brownian affinity gating model, which predicts that the NPC possesses an aqueous central channel that is occluded by clouds of FG repeats on both its nuclear and cytoplasmic openings (Rout et al., 2000). The suggested FG repeat clouds are considered a direct consequence of the Brownian motion of many individual end-grafted FG repeat bristles, hence the name (Rout et al., 2000). Notably, this particular arrangement of FG repeats was proposed based on the finding that antibody-coated gold particles directed against FG repeats exclusively localized to the peripheries of the NPC in the yeast *Saccharomyces cerevisiae* (Rout et al., 2000). Artificial nanopores analogously surface-coated with FG repeats indeed showed selectivity for NTRs in contrast to inert molecules (Jovanovic-Talisman et al., 2008; Kowalczyk et al., 2011), in agreement with computational models of NPCs with such a FG repeat topology (Moussavi-Baygi et al., 2011).

However, based on the available structural information on the architecture of the NPC, including the knowledge of defined FG repeat anchoring sites, it is now undisputed that the central channel itself is densely crowded with FG repeats (see Section 1.2 for details). Moreover, it has been explicitly shown that the FG repeats localized to both the cytoplasmic and nuclear face of the NPC are dispensable for selective nuclear transport (Strawn et al., 2004), highlighting that the Brownian affinity gating model is an oversimplification of the actual *in vivo* situation unable to reliably explain the selectivity properties of NPCs.

1.3.2 Polymer brushes and the reversible collapse model

Notwithstanding the evidence against Brownian affinity gating, the concept of virtual gating has not been abandoned, but rather conjoined with alternative FG repeat arrangements, in particular entangled polymer brushes (see below). In materials science, polymer brushes describe a dynamic conformation that long-chain polymers attached to surfaces at high densities will occupy in an effort to compensate for the free-energy costs of their crowding (Milner, 1991). By constantly exploring a large number of possible conformations, individual polymers try to stretch away from their neighbors, in turn considerably extending from the surface into space. Notably, isolated polymers (especially polypeptides) free in solution would rather adopt much more compact structures, illustrating that polymer brushes are under severe 'entropic strain'. Inert macromolecules are consequently excluded from polymer brushes when the additional decrease in entropy (of both the brush and the molecule) resulting from entering cannot be further compensated (Bright et al., 2001). The degree of 'entropic strain' and (inversely related) the thickness of a polymer brush are strongly influenced by the length and grafting density of its constituents (amongst other parameters; see also below).

Evidence that FG repeats can indeed form such polymer brushes comes from several studies investigating the biophysical properties of surface-immobilized FG repeats (Lim et al., 2007; Lim and Deng, 2009; Eisele et al., 2010) and related molecular dynamics simulations (Miao and Schulten, 2009; 2010). Particularly interesting is a series of studies describing the 'collapse' of FG repeat polymer brushes upon NTR binding (Lim et al., 2007). Applied to the NPC, this finding has been interpreted to mean that binding of NTRs causes a contraction of the FG repeats towards their anchor points, thereby mechanically drawing a translocating species deep into the nuclear pore. Next, via a series of binding-collapsing and unbinding-extending steps, NTR•cargo complexes are supposed to translocate through the pore. Finally, either binding of RanGTP in the nucleus or aided GTP hydrolysis in the cytoplasm most likely terminate the translocation process and confer directionality. Taken together, this 'reversible collapse' model thus not only involves a virtual gating mechanism explaining the rejection of inert molecules, but also suggests a mechanical propelling of the translocating species resulting from morphologically changes of the permeability barrier (Lim et al., 2007).

However, given the high cellular concentration of NTRs (>10 µM) and the great number of translocations per NPC (~1000 a second), this model fails to explain why the polymer brush is not constantly collapsed. Apart from such conceptual objections, the experimental setup chosen here was itself biased.

First, in the study described above (Lim et al., 2007), the FG repeats were immobilized via terminal cysteine residues to a gold surface, which however had not been quenched after coupling. Hence, it is likely that the NTR used (i.e. Impβ, which contains multiple endogenous cysteines) unintentionally bound to the gold surface as well, in turn causing the collapse of the FG repeats.

Second, the grafting densities of FG repeats were considerably lower than expected in the nuclear pore, thus *a priori* favoring compact states that can easily be pulled further together by multivalent NTRs. In fact, no such drastic morphological changes have been reported even in response to massive NTR binding (resulting in a ~45% increase of total protein mass) in a similar study with FG repeats surface-immobilized at a density comparable to the yeast NPC environment (Eisele et al., 2010).

1.3.3 The reduction of dimensionality model

The reduction of dimensionality models attempts to explain how partially collapsed polymer brushes can still function as a selectivity barriers in nucleocytoplasmic transport. It assumes that the subsided FG repeats form a coherent layer of continuous NTR binding sites along the inner walls of the NPC in response to steady translocation activity, concomitantly envisioning facilitated nuclear pore passage as a two-dimensional random walk ('sliding') instead of a three-dimensional diffusion process as in other models (Peters, 2005; 2009). Non-binding molecules however cannot benefit from this 'reduction of dimensionality' and are thus limited to diffusion through the central channel. To explain the size selectivity of NPCs, the central channel is thought to be clogged by the hydrophilic spacer regions of the FG repeats (which do not contain NTR binding sites and are thus supposedly not collapsed to the channel walls), leaving only a narrow tube available for unhindered transit of small objects.

However, it is not plausible how fully assembled ribosomal subunits with a diameter of ~25nm should be able to cross such strait channels. The model thus essentially fails to explain the known selectivity of the NPC permeability barrier for molecules with sizes of different orders of magnitude.

1.3.4 The selective phase model

In contrast to the previously described views, the selective phase model assumes an additional property of FG repeats, namely their ability to interact with each other (Ribbeck and Görlich, 2001). According to the the model, the hydrophobic patches of the FG repeats (i.e. the FG motifs also serving as NTR binding sites) associate and thereby crosslink the entire domains into a sieve-like meshwork, the selective phase (Ribbeck and Görlich, 2001; 2002). Molecules above the mesh size can only partition into the selective phase when they are able to open individual meshes. In view of the model, the multivalent NTRs achieve this by transiently and locally binding to multiple interacting FG motifs, thus becoming a dynamic part of the meshwork capable of diffusing within the selective phase (Ribbeck and Görlich, 2001; 2002).

Remarkably, purified yeast and vertebrate FG repeats can indeed form a selective phase *in vitro*, namely macroscopic hydrogels that are molecular sieves mimicking the passive permeability properties of NPCs (Frey et al., 2006; Frey and Görlich, 2007; 2009; Milles and Lemke, 2011). In theory, the restrictive potential of FG hydrogels towards passive fluxes of inert molecules should largely depend on the availability of interaction partners for the formation of meshes. Allowedly, the selectivity properties of NPCs can only be matched by FG hydrogels when the concentration of FG repeat units is high enough to ensure that all of them have the chance to find a sufficiently close inter-repeat binding partner (Frey and Görlich, 2007). This threshold concentration is called the 'saturation limit'. Strikingly, saturated FG hydrogels also reproduce the fundamental aspects of facilitated NPC passage: rapid entry of NTR•cargo complexes into the permeability barrier, intragel diffusion at rates that are in agreement with observed NPC passage times for such complexes, and exit of the complexes from the gel (Frey and Görlich, 2007; 2009; Milles and Lemke, 2011).

The importance of hydrophobic clusters for hydrogel formation is illustrated by the finding that mutant repeats lacking FG motifs cannot form hydrogels anymore (Frey et al., 2006). Likewise, the permeability barrier of NPCs breaks down *in situ* when hydrophobic contacts are interrupted with low concentrations of aliphatic alcohols such as (cyclo-) hexanediol (Ribbeck and Görlich, 2002). This highlights the crucial importance of hydrophobic interactions between FG repeats in both the model system and the nuclear pore. Additional functional arguments for the presence of a selective phase *in vivo* are the good agreement of the passive diffusion channel sizes observed in the hydrogel system and nuclear pores (in contrast to the other models, as discussed above) (Frey and Görlich, 2009; Mohr et al., 2009) and the direct morphological evidence for a hydrophobic meshwork in the central channel of authentic NPCs (Kramer et al., 2008). The findings that polar cargoes exert a significant 'drag' on NTRs during NPC passage and that this effect can be overcome by binding of additional NTRs, despite the successive increase in size, have been interpreted such that NTRs are crucially required to confer solubility in the hydrophobic selective phase (Ribbeck and Görlich, 2002; Lowe et al., 2010).

Most critics of the selective phase model question the existence of a proteinaceous sieve-like meshwork in the nuclear pore based on the argument that hydrogels are formed under non-physiological conditions and thus resemble an artifact (Weis, 2007). However, with the recent finding that only cohesive FG repeats are capable of forming functional permeability barriers in a physiological context, direct compelling evidence for the selective phase model was now provided (Hülsmann et al., 2012).

1.3.5 The forest model

Like the selective phase model, the recently proposed two-gate or forest model (Patel et al., 2007; Yamada et al., 2010) accommodates the finding that FG repeats are cohesive. However, it takes into consideration that this does not apply to all yeast FG repeats (Patel et al., 2007). Indeed, when different FG domains within individual FG repeats were independently analyzed with a focus on the confirmations they adopt (Yamada et al., 2010), two major kinds were identified: cohesive FG domains that adopt a globular, collapsed-coil configuration vs. highly charged, non-cohesive FG domains in extended-coil configurations. Notably, some FG repeats contained FG domains of both types in a bimodal distribution along their polypeptide chains.

In view of the forest model, cohesive FG repeats comprising only collapsed-coil FG domains are thought to 'fold' back onto the channel walls of the NPC to form globular structures that are reminiscent of 'shrubs'. In contrast, bimodal FG repeats containing both non-cohesive FG domains and terminal cohesive FG domains are proposed to form 'trees', with the non-cohesive parts forming extended-coils that 'stem' away the cohesive FG domains from their anchor sites. As a consequence, two major transport zones are expected, either encompassing the space in-between the 'shrubs' and the 'treetops', or the central channel left behind by the non-overlapping, yet laterally interacting 'treetops' (Yamada et al., 2010).

In a way, the forest model thus seems to have emerged from a 'melting pot' of all previously available models and interpretations. But despite the convincing evidence that FG repeats contain different FG domains (see also Ader et al., 2010 and Sections 2.2), such a high degree of self-organization into a defined, symmetric structure as suggested here for the intrinsically disordered and flexible FG repeats seems rather implausible. In contrast to the predictions of the forest model, the only partially cohesive FG repeats from Nsp1 indeed fail to form a permeability barrier in WGA-depleted NPCs (Hülsmann et al., 2012).

1.3.6 At the watershed: do FG repeats interact?

To conclude, it is clear from the different models and experimental evidence that both the nature and function of the NPC permeability barrier critically depend on the degree to which the highly crowded FG repeats interact within the confined space of the nuclear pore. In this regard, it is of particular interest whether cohesiveness, like the ability to bind NTRs, is an essential feature of FG repeats or rather a meaningless detail observed in only a few yeast and vertebrate species.

The central argument of this work is that if a selective phase was indeed the essential feature of the NPC permeability barrier, then the cohesiveness of FG repeats should be evolutionary conserved in all Eukaryotes. Hence, **the primary aim of this study was to explore the potential of FG repeats from all eukaryotic clades to form selective hydrogels mimicking the permeability properties of NPCs.**

1.4 The Relationship between FG Repeats and NQ-rich Prions

So far, it was assumed that only the hydrophobic NTR-binding sites within FG repeats (i.e. the FG motifs) are functionally relevant, both in regard to NTR binding and formation of the permeability barrier (see previous section for detailed discussion). Recent solid-state NMR data however suggested that also the intervening spacer regions play a key role for hydrogel formation and stability (Ader et al., 2010). In particular, the asparagine (N)-, glutamine (Q)- and threonine (T)-rich spacers of the N-terminal FG repeat region from the yeast nucleoporin Nsp1 were found to readily assemble into interchain β-sheets, thereby providing a plausible structural explanation for their sticky nature. Likewise, molecular dynamics simulations showed that initially unstructured short model FG repeat peptides derived from Nsp1 progressively aggregate into higher order clusters, which are dependent on hydrophobic contacts, but yet mainly stabilized by backbone-backbone hydrogen bonds between spacer residues of emerging adjacent β-strands (Dölker et al., 2010). The decreased propensity of a N/Q->G mutant version of the N-terminal Nsp1 FG repeats to form hydrogels verified that asparagine and glutamine indeed promote cohesiveness (Steffen Frey, unpublished data). In general, a strong correlation between cohesiveness and NQ-content is evident for yeast FG repeats (Patel et al., 2007; Frey and Görlich, 2009; Ader et al., 2010; moreover Koray Kirli and Michael Ridders, personal communication). Thus, in sharp contrast to the plethora of models *per se* neglecting inter-FG repeat contacts in their explanation of NPC function (see previous Section), the spacers separating single FG motifs can clearly provide additional means to confer cohesiveness, thereby severely influencing the structures FG repeats will form within the NPC and hence also the selectivity mechanism underlying nucleocytoplasmic transport.

Remarkably, interchain β-sheets are also the defining structural hallmark of amyloids (Sunde and Blake, 1997). The latter are ordered, fibrillar aggregates formed by a number of different proteins with diverse cellular functions (Chiti and Dobson, 2006). Extensive intermolecular interactions, predominantly hydrogen bonding between the protein backbones of laterally opposing monomers, but also stacking interactions between compatible side chains of aligned β-sheets, stabilize the pseudocrystalline fibers (Lührs et al., 2005; Nelson et al., 2005; van der Wel et al., 2007; Shewmaker et al., 2011). Interestingly, despite their proteinaceous nature, amyloid aggregates were originally mistaken to consist of starch, as also they can be stained by iodine, hence explaining their misleading name (Kyle, 2001).

Amyloid fibrils may emerge spontaneously and are usually the result of protein mis-folding, which causes conventionally folded domains or even full length proteins to aggregate (Chiti and Dobson, 2006). However, amyloid formation can also be induced by prions, transmissible infectious proteins which are thought to trap newly synthesized or existing protein monomers in the growing fibrils by preventing their correct folding or causing their (partial) unfolding (Chiti and Dobson, 2006; Halfmann et al., 2010). Note that the term 'prion' is more generally also used to denote proteins that have the capacity to form amyloids in a self-templating manner (Alberti et al., 2009). Given the abnormal conformation of the amyloid form, it is not surprising that a number of pathological conditions, such as Alzheimer's disease or lysozyme amyloidoses, are associated with amyloid aggregates (Chiti and Dobson, 2006). Yet, more and more studies suggest that amyloids also play a role in normal cellular function, e.g. in the antiviral immunse response of mammals (Hou et al., 2011) or long-term memory formation in sea slugs and flies (Si et al., 2010; Majumdar et al., 2012). Another noteworthy example for the natural (extracellular) use of amyloid-forming proteins is fibroin, the major structural constituent of silk. Fibroin forms the amyloid-like β-sheet 'nanocomposites' conferring extreme (tensile) strength to e.g. the cocoons of the silkworm or silk-based textiles (see e.g. Kushner and Guan, 2011).

Notably, it is well know that NQ-rich sequences can, despite their general tendency to remain disordered on a monomer level (Weathers et al., 2004; Toombs et al., 2010), spontaneously self-assemble into amyloids (Perutz et al., 2002; López de la Paz and Serrano, 2004; Alberti et al., 2009). Taken together, the striking similarities between NQ-rich FG repeats and prions on a sequence and structural level pose the question whether such prions can also form selective hydrogel-based permeability barriers. Accordingly, **the second aim of this study was to elucidate the sequence features required for functional permeability barriers by gradually converting a *bona fide* prion into a FG repeat.**

NQ-rich prions have been most extensively characterized in the yeast *Saccharomyces cerevisiae*, in which such proteins are remarkably abundant (Michelitsch and Weissman, 2000; Harrison and Gerstein, 2003). Hence, the prototypical prion-determining domain (PrD) of the yeast translation-termination factor Sup35 was chosen as a model NQ-rich prion to address the aforementioned objectives. Sup35 is arguably the most extensively studied prion in the literature and another example for non-pathogenic implications of amyloids (see below). Interestingly, the PrD is in principle dispensable for the normal cellular function of Sup35. With a frequency of 1 in 10 however (Lancaster et al., 2010), any given PrD adopts an self-perpetuating amyloid conformation, causing more and more PrDs of additional Sup35 molecules to integrate into the growing fiber. Eventually, nearly the entire pool of soluble Sup35 molecules is sequestered in the insoluble amyloid form (Patino et al., 1996). As a consequence, the translation termination activity is markedly reduced and the rate of stop codon read-through concomitantly boosted. However, this is thought to give rise to a variety of new traits emerging from the previously cryptic genetic variation, which might be beneficial under certain circumstances (True and Lindquist, 2000; Halfmann et al., 2012a). In view of the model, by the time a yeast colony has reached appreciably size, Sup35 will have switched into the amyloid conformation in a few cells. If this in turn causes harm to the cells, only a few individuals of a large population may have to be sacrificed. In contrast, if the environment favors the new phenotype, survival of at least some individuals can be ensured when the majority of the population must otherwise perish. Due to the amyloid-severing function of Hsp104 (Chernoff et al., 1995; Kryndushkin et al., 2003), the self-sustaining amyloid state can be passed on to daughter cells, thus also allowing beneficial phenotypes to be epigenetically inherited. Sporadically, the amyloid state may be lost and the original phenotype becomes available again.

1.5 The Function of FG Repeats in Nuclear Dualism

Notably, selective exchange between nucleus and cytoplasm encompasses an additional level of complexity in Ciliates such as *Tetrahymena thermophila*. These unicellular eukaryotes contain two morphologically and functionally different nuclei in a common cytoplasm, a peculiarity referred to as nuclear dualism. The 'somatic' macronucleus (MAC) contains multiple, highly rearranged copies of the genome and is the site of all gene transcription, the latter of which is comparable in extent to the expressed gene content of mammals. In contrast, the diploid micronucleus (MIC) is transcriptionally silent during vegetative growth and serves as a germ-line equivalent. As deduced below, nuclear dualism necessitates selective and reliable transport of the correct cargoes into and out of the right nucleus in order to maintain the operational integrity of the cell at all times. Yet despite the great efforts such a sophisticated cellular organizations requires, the unique combination of a highly dynamic and 'innovative', as well as a safely stored 'backup' genome is arguably one of the major reasons for the success of the rather early branching, free-living and (theoretically) immortal ciliated protozoans to adapt to and thrive in ever changing environmental conditions.

1.5.1 The complex life cycle and unique molecular biology of *Tetrahymena* rely on extensive nucleocytoplasmic exchange

The *Tetrahymena* life cycle features to two different stages: asexual reproduction via binary fission and the non-reproductive exchange of genetic information between mating-competent cells, a process known as conjugation (Figure 1.6). Not surprisingly, the two nuclei play fundamentally different roles and have dramatically different fates in either process.

During asexual multiplication, the MAC divides amitotically with a concomitant random segregation of the genetic information, whereas the MIC undergoes an abbreviated, but otherwise normal mitosis. The nuclear envelopes of both nuclei stray intact throughout the two events, which strikingly occur asynchronously. Hence, constant nucleocytoplasmic transport activities are required by both nuclei to meet their specific demands. A prominent example for differential MAC and MIC protein compositions are linker histones, which organize chromatin into higher order structure. MACs contain prototypical H1-type linker histones, which are extremely basic and subject to many of the known common post-translational histone modifications (Gorovsky et al., 1974; Johmann and Gorovsky, 1976). In contrast, the MIC chromatin is condensed by four different unusual linker proteins collectively called micronuclear linker histones (MLHs) (Shen et al., 1995), which originate from a polyprotein precursor after proteolytic processing (Wu et al., 1994). Another MIC-specific histone variant is CNA1, a component of the centromere (Cervantes et al., 2006). This finding is in good agreement with the fact that faithful chromosome segregation driven by a mitotic spindle happens only in the MIC, but never in the MAC. Notably, the study of histones in *Tetrahymena* also provided the first proof for the presence of selective and nucleus-specific transport mechanisms almost 25 years ago (White et al., 1989). Taken together, the vegetative life cycle allows *Tetrahymena* cells to flourish rapidly, with doubling times well below two hours, despite their large size and complexity (Cassidy-Hanley, 2012).

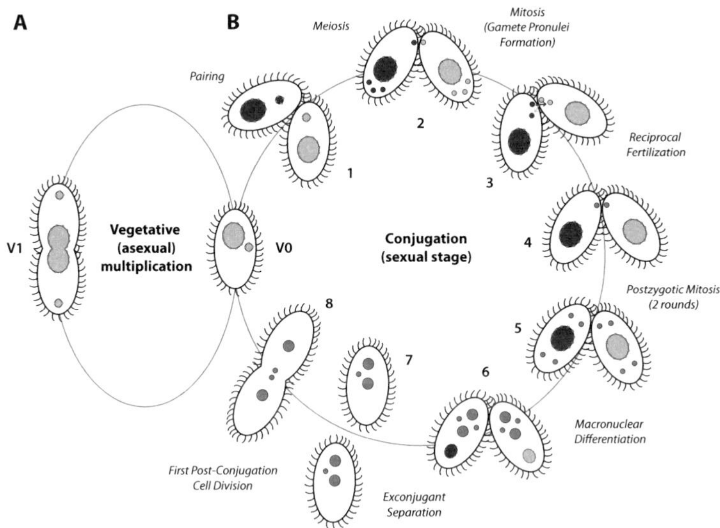

Figure 1.4. The life cycle of *Tetrahymena thermophila*. Adapted from (Orias et al., 2011). (A) Vegetative (asexual) reproduction. V0 depicts a vegetative cell and V1 a cell undergoing binary fission. Please note that the MAC divides amitotically, whereas the MIC divides via closed mitosis. The nuclear envelopes stay intact throughout cell division. (B) Conjugation, the sexual (non-reproductive) stage of the *Tetrahymena* life cycle. Note that all following steps occur synchronously in both conjugants. 1. Fusion of starved cells with opposite mating types. 2. The micronuclei undergo meiosis in both conjugants, yielding four haploid nuclei, three of which are destroyed again by programmed nuclear death (PND). 3. The remaining nucleus divides mitotically to give rise to two gamete pronuclei. One of the pronuclei from each partner cell is reciprocally exchanged. 4. The incoming migratory pronucleus fuses with the remaining stationary pronucleus to form a diploid zygotic nucleus in each conjugant. 5. The zygotic nucleus undergoes two rounds of mitosis to generate four diploid anlagen, which eventually differentiate into two new polyploidy MACs and two new diploid MICs. 6. The new MACs and MICs are aligned in the cells, whereas the parental MAC is resorbed via PND. Note that this period is characterized by extensive information exchange between the developing nuclei and the parental MAC (see Section 1.5 for details). 7. Exconjugants separate and one new MIC is broken down by PND. 8. The first post-conjugation cell division is required to

restore the vegetative state. The two new MACs, as well as a mitotic daughter of the surviving new MIC, are equally distributed between the daughter cells. Strictly seen, this step is not part of the conjugation process. For clarity, the first post-conjugation cell division is only depicted for one of the two exconjugants.

Upon environmental stress however, two compatible *Tetrahymena* cells may temporarily fuse and proceed with conjugation (Figure 1.6). In this process, the MICs of both participating cells first undergo meiosis to yield four haploid gametes, three of which are immediately degraded, as typical for asymmetric 'female' meiosis. The remaining haploid nuclei divide mitotically, generating a migratory and a stationary pronucleus in each cell. The migratory pronuclei are reciprocally exchanged between the two conjugants and fuse with the stationary pronuclei of the opposite mating partner to give rise to diploid zygotic nuclei. Each zygote eventually endures two rounds of mitosis, providing four diploid anlagen, two of which will differentiate into new MACs and two of which will form new MICs.

Remarkably, massive genome adjustments occur in the course of MAC differentiation (reviewed in Karrer, 2012): (i) the chromosomes are broken up into manifold pieces and new telomers have to be added, (ii) the single-copy ribosomal RNA gene is converted into a giant palindrome and amplified to about 10,000 copies, and (iii) large portions of the genome (referred to as internal eliminated sequences, IES) are site-specifically deleted in an RNAi-mediated process, while simultaneously, (iv) multiple rounds of DNA replication lead to a highly polyploid macronuclear genome. At this point, the parental MACs are still present, as they are required to assist in IES processing in the differentiating MACs.

In the current view (reviewed in Chalker, 2008 and Karrer, 2012), the micronuclear genome, which is silence most of the time, is now bidirectionally transcribed. The resulting transcripts are processed to produce small, double-stranded 'scan' RNAs (scRNA) covering the complete micronuclear genome. After export to the cytoplasm and formation of specialized RISC (RNA induced silencing complex) species, the scRNA-RNPs are transported to the parental MAC, where the macronuclear genome is scanned and 'compared' to the micronuclear genome. All complexes containing scRNA bearing no homology to macronuclear sequences are kept, whereas all others are degraded. Thus, via this process, the IES elements are defined *de novo* every time cells undergo conjugation, rather than being permanently encoded in the genome. The sorted and approved scRNA-RNPs are eventually transferred to the developing MAC anlagen, where they direct the excision and subsequent degradation of the IES elements. However, after this process is completed, the parental MACs are completely degraded via massive autophagy and the still paired conjugants finally split up.

Once separated, one of the new MICs is likewise degraded in both cells. Yet, the original vegetative state is only restored after the first vegetative cell division after conjugation, during which one of the two new MACs and a copy of the mitotically duplicated MIC are distributed to the daughter cells. Again, the nuclear envelopes of all nuclei stay intact throughout every event outlined above.

In summary, up to five nuclei temporarily co-exist in the course of conjugation, of which all deliver or demand a different subset of specific factors to differentiate and uphold cellular function. Likewise, two distinct nuclei need to be coordinated at all times in the vegetative state. Hence, there is an obvious crucial need for the spatially and temporally fine-regulated distribution of macromolecules to the correct nuclei.

1.5.2 Nucleus-specific nucleocytoplasmic exchange in *Tetrahymena*

However, nucleus-specific nuclear transport cannot solely rely on exclusive transport factors and unambiguous targeting signals on the cargoes, owing to the fact that they all periodically converge in the common cytoplasm and hence will face both type of nuclei as a result of the translocation process. Thus, different kinds of NPC permeability barriers with orthogonal specificities are first and foremost required to guard the MAC and MIC. Evidence for this notion was provided with the identification of four different Nup98 paralogs in *Tetrahymena*, of which one pair exclusively localizes to the MAC and the other to the MIC (Malone et al., 2008; Iwamoto et al., 2009). So far, these are the only know nucleus-specific Nups involved in forming the MAC or MIC permeability barrier, respectively (see also Table 1.3). Yet, Nup98 was recently shown to govern the characteristics of the metazoan nuclear pore (Hülsmann et al., 2012), suggesting that the fundamentally different *Tetrahymena* Nup98 paralogs are nevertheless likely to play a pivotal role in establishing the different specificities of the MAC and MIC permeability barriers. Indeed, when the macro- and micronuclear FG repeats were exchanged, the resulting chimeric NPCs neither allowed the import of a MAC-, nor a MIC-specific cargo anymore (Iwamoto et al., 2009), indicating that Nup98 can at least efficiently suppress the passage of wrongly destined transport complexes in *Tetrahymena*.

Table 1.3. *Known components of the* Tetrahymena *nuclear transport apparatuses.*

\- Ran System \-				
Protein Name	Original Name	Localization	Identification	Reference
TtRan	Ran1	MAC	Cloning	(NAGATA et al., 1994; Liang et al., 2012)
\- Scaffold Nups \-				
Protein Name	Original Name	Localization	Identification	Reference
Seh1	-	MAC/MIC	*In silico*	(Iwamoto et al., 2009)
Sec13	-	MAC/MIC	*In silico*	(Iwamoto et al., 2009)
gp210	-	MAC/MIC	*In silico*	(Iwamoto et al., 2009)
TtNup93	-	MAC/MIC	*In silico*	(Iwamoto et al., 2009)
TtNup96	-	MAC/MIC	*In silico*	(Iwamoto et al., 2009)
TtNup155	-	MAC/MIC	*In silico*	(Iwamoto et al., 2009)
\- FG Nups \-				
Protein Name	Original Name	Localization	Identification	Reference
TtNup50	Nup1	MAC/MIC	*In silico*	(Malone et al., 2008), (Iwamoto et al., 2009)

Protein Name	Original Name	Localization	Identification	Reference
TtNup54	-	MAC/MIC	*In silico*	(Iwamoto et al., 2009)
TtMacNup98A	-	MAC	*In silico*	(Iwamoto et al., 2009)
TtMacNup98B	Nup3	MAC	*In silico*	(Malone et al., 2008), (Iwamoto et al., 2009)
TtMicNup98A	Nup4	MIC	*In silico*	(Malone et al., 2008), (Iwamoto et al., 2009)
TtMicNup98B	-	MIC	*In silico*	(Iwamoto et al., 2009)
TtNup308	-	MAC/MIC	*In silico*	(Iwamoto et al., 2009)
– Nuclear Transport Receptors (NTRs) –				
Protein Name	Original Name	Localization	Identification	Reference
TtImpB1	IMB1	MAC/MIC	*In silico*	(Malone et al., 2008)
TtImpB2	IMB2	cytoplasmic	*In silico*	(Malone et al., 2008)
TtImpB3	IMB3	MAC only	*In silico*	(Malone et al., 2008)
TtImpB4	IMB4	MAC only	*In silico*	(Malone et al., 2008)
TtImpB5	IMB5	MAC/MIC	*In silico*	(Malone et al., 2008)
TtImpB6	IMB6	MIC preferred	*In silico*	(Malone et al., 2008)
TtImpB7	IMB7	MAC/MIC	*In silico*	(Malone et al., 2008)
TtImpB8	IMB8	MAC/MIC	*In silico*	(Malone et al., 2008)
TtImpB11	IMB11	cytoplasmic	*In silico*	(Malone et al., 2008)
TtCrm1	XPO1	MAC preferred	*In silico*	(Malone et al., 2008)
– Import Adaptors –				
Protein Name	Original Name	Localization	Identification	Reference
TtImpA1	IMA1	MAC	*In silico*	(Malone et al., 2008)
TtImpA2	IMA2	MIC	*In silico*	(Malone et al., 2008)
TtImpA3	IMA3	MIC	*In silico*	(Malone et al., 2008)
TtImpA4	IMA4	MIC	*In silico*	(Malone et al., 2008)
TtImpA5	IMA5	MIC	*In silico*	(Malone et al., 2008)
TtImpA6	IMA6	MAC	*In silico*	(Malone et al., 2008)
TtImpA8	IMA8	MIC*	*In silico*	(Malone et al., 2008)
TtImpA10	IMA10	MIC	*In silico*	(Malone et al., 2008)
TtImpA11	IMA11	MIC	*In silico*	(Malone et al., 2008)
TtImpA12	IMA12	MIC*	*In silico*	(Malone et al., 2008)
TtImpA13	IMA13	MIC	*In silico*	(Malone et al., 2008)

* highly enriched

Surprisingly though, all efforts to identify nucleus-specific NTRs so far revealed that apart from two MAC-specific NTRs, all others localize to both the MAC and MIC (Malone et al., 2008), as summarized in Table 1.3. Maybe even more startling, *Tetrahymena* supposedly contains at least ten different Importin α-like import adaptors (see also Table 1.3), of which most specifically localize to the MIC, whereas only one strongly associates with the MAC (Malone et al., 2008). For comparison, yeast contains only one and humans six Imp-α family members (Mason et al., 2009).

Taken together, the unique cell biology of ciliates and the initial discoveries of the first components of the *Tetrahymena* nuclear transport machineries present us with an excellent opportunity to investigate the influence of different FG repeats and their (sequence) features on the formation of permeability barriers with opposite specificities, thereby advancing our understanding of both nuclear dualism and the fundamental selectivity/translocation mechanism of the NPC permeability barrier. Therefore, **the third aim of this study is to reconstitute nuclear dualism in the hydrogel model system.**

2. Results

2.1 The cohesiveness of FG repeats appears to be evolutionary conserved in all eukaryotic clades.

The NPC is guarded by a permeability barrier that restricts the free diffusion of molecules in a size-dependent manner. It is constituted of FG repeats, yet the question how FG repeats are physically arranged in the nuclear pore to allow for the phenomena of translocation and selectivity remain highly debated.

Arguably the most plausible hypothesis is the selective phase model (see Section 1.3), which suggests that the non-globular FG repeat domains filling the nuclear pore heterotypically interact to form a three-dimensional sieve-like meshwork, the selective phase (Ribbeck and Görlich, 2001; 2002). Molecules above the mesh size can only partition into and diffuse within the selective phase when they are able to open individual meshes.

Accumulating evidence indeed points to such a cohesive nature of FG repeats (Frey et al., 2006; Frey and Görlich, 2007; Patel et al., 2007; Frey and Görlich, 2009; Yamada et al., 2010; Milles and Lemke, 2011 and A. Labohka, personal communication), which notably enables them to form a selective phase *in vitro*: macroscopic hydrogels that reproduce the fundamental aspects of nucleocytoplasmic transport.

The selective phase model predicts that every functional NPC needs at least one type of cohesive FG repeat allowing for the formation of the selective phase. Thus, if the NPC permeability barrier were indeed a selective phase, then the cohesiveness of FG repeats and their ability to form hydrogels reproducing the fundamental properties of the NPC permeability barrier should be evolutionary conserved in all Eukaryotes.

2.1.1 Identification of FG repeats in the domain Eukarya.

To challenge this assumption, FG repeats first needed to be systematically identified in each of the six proposed eukaryotic clades (Simpson and Roger, 2004; Keeling et al., 2005; Lane and Archibald, 2008). As noted before (Bapteste et al., 2005; Denning and Rexach, 2007), this is not a trivial objective, as recognition and assignment of proteins typically relies on evolutionary kinship, which is generally detected based on primary sequence and/or protein structure similarity. FG repeats however are (i) intrinsically disordered and hence even lack secondary structure elements, (ii) contain few to many FG motifs of different flavors (e.g. FG, FxFG or GLFG motifs) in irregular intervals and (iii) vary significantly in the length and amino acid composition of the spacer regions separating individual FG clusters.

A notable exception is Nup98, a FG Nup located to the central channel of (vertebrate) NPCs and the most crucial contributor to the (vertebrate) permeability barrier (Hülsmann et al., 2012). It is expressed as a fusion pre-protein with the NPC scaffold component Nup96 (see Figure 2.1.A) and contains a highly conserved, folded nucleoporin2 autopeptidase domain flanking its FG repeat region (see Figure 2.1.B and Hodel et al., 2002). This domain is required to auto-catalytically process the pre-protein into the mature versions of Nup98 and Nup96 (see Figure 2.1.A and Fontoura et al., 1999). It moreover participates in anchoring mature Nup98 to the NPC after processing (Hodel et al., 2002; Ratner et al., 2007). Notably, Nup98 was the only FG Nup that was reliably and reproducibly identified in all previous comparative genome studies of the NPC (Mans et al., 2004; Bapteste et al., 2005; Neumann et al., 2010).

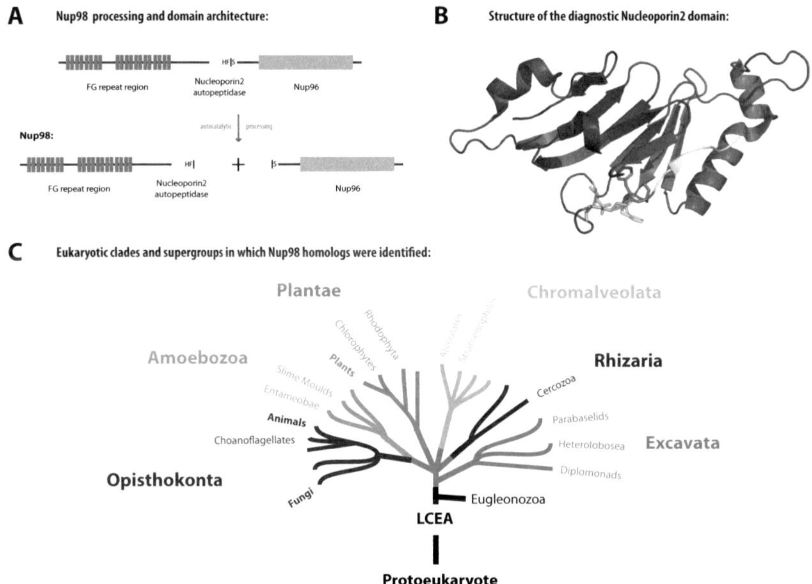

Figure 2.1. Identification of Nup98 homologs in Eukarya. (A) In most eukaryotes, the integral permeability barrier component Nup98 is expressed as a fusion protein with the NPC scaffold module Nup96. As a result the autocatalytic activity of the Nucleoporin2 domain, the two parts of the fusion protein are post-translationally cleaved at a defined site (i.e. the amino acid motif HF|S, where | denotes the peptide bond cleaved). Note that after cleavage, the Nucleoporin2 domain further serves to anchor Nup98 to the NPC via interactions with Nup96 (Ratner et al., 2007). (B) The structure of the Nucleoporin2 domain was solved (Hodel et al., 2002) and recognized as highly conserved. In the present study, it was hence used as a diagnostic marker to identify Nup98 homologs in Eukarya. (C) Shown here are the eukaryotic clades and the supergroups in which Nup98 homologs have been identified. The tree topology is based on (Simpson and Roger, 2004; Keeling et al., 2005; Adl et al., 2007; Cavalier-Smith, 2010a).

In the present study, the nucleoporin2 domain was used as a diagnostic feature to systematically search the non-redundant NCBI protein database for homologs of Nup98 in every eukaryotic clade on a supergroup level with unprecedented completeness to create a custom database of Nup98-derived FG repeats. A graphical summary of the supergroups in which Nup98 homologs were found is given in Figure 2.1.C. The FG repeat regions were extracted semi-automatically from the full-length sequences based on the occurrence of *bona fide* FG and similar hydrophobic F-rich motifs, as well as disorder predictions (see Methods for details). In total, the database contained more than 250 entries, covering the complete eukaryotic biodiversity, including for the first time also FG repeats from the kingdom Rhizaria (see Figure 2.1.C).

2.1.2 Sequence analysis of the Nup98-derived FG repeats.

To characterize the Nup98-derived FG repeats, it was first analyzed how they differ from folded, globular proteins. Therefore, the average amino acid compositions of all identified FG repeats and a pool of 43,634 globular proteins from the PDB were compared (Figure 2.2.A).

Interestingly, the Nup98-derived FG repeats were found to be (i) highly enriched (i.e. >2-fold abundance) in glycine, threonine and phenylalanine and (ii) markedly enriched (i.e. <2-fold higher prevalence) in serine, asparagine and glutamine residues. They (iii) contain comparable amounts of alanine and proline, but are (iv) very poor (i.e. up to 10-fold less enriched in some cases) in charged residues and (v) all aromatic and hydrophobic residues except for phenylalanine.

When considering the overall hydrophobicity (as determined based on the scale compiled by Fauchere and Pliska, 1983), no striking differences between the Nup98-derived FG repeats and the globular proteins were detected (Figure 2.2.B). This is an interesting finding given that globular proteins tend to 'burry' most of their hydrophobic residues in the core of their folds, whereas the hydrophobic FG motifs of FG repeats need to remain accessible for NTRs.

Furthermore, the FG repeats carry a considerable overall positive net charge, albeit their low content in aspartate, glutamate, arginine and lysine (DERK) residues. The proteins from the PDB in contrast are slightly negatively charged, yet with a very broad distribution around the mean (Figure 2.2.B).

Taken together, apart from containing FG motifs, the Nup98-derived FG repeats were found to be characteristic of (i) a high content of glycine, threonine, serine, asparagine, glutamine and alanine residues, (ii) a strong bias of phenylalanine over any other hydrophobic and/or aromatic residue and (iii) a low charge density with an overall positive charge.

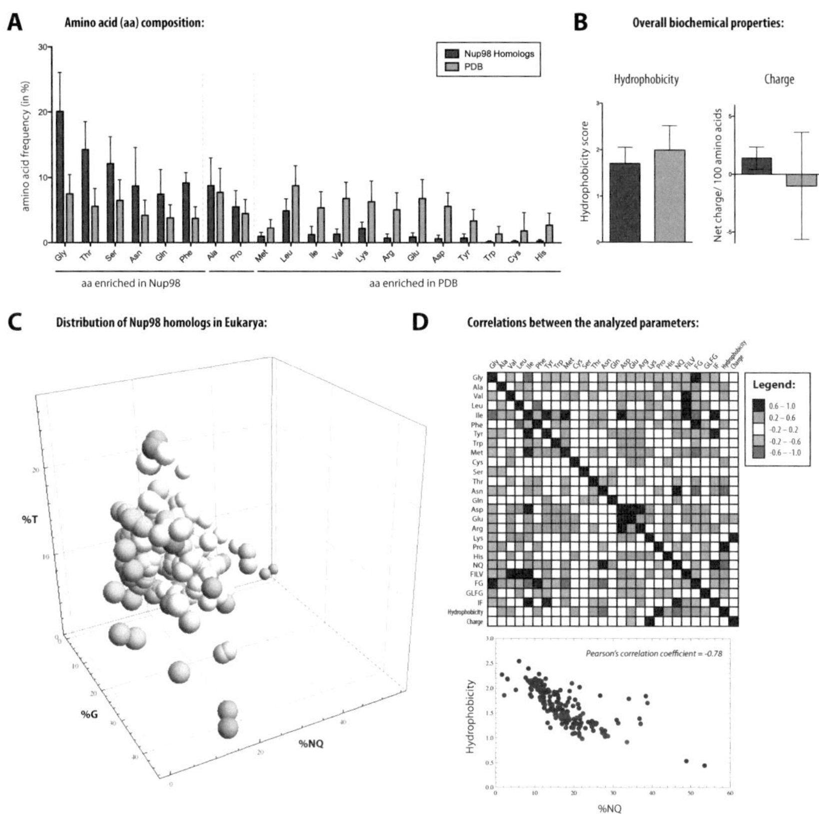

Figure 2.2. Sequence analysis of the Nup98-derived FG repeats. The amino acid sequences of the >250 identified non-globular FG repeat domains extracted from the Nup98 homologs were compared to 43,634 globular proteins from the PDB. Given in (A) are the mean frequencies of all 20 standard amino acids (and the corresponding standard deviations) in the FG repeats and the globular proteins, grouped according to their enrichment in either protein category. (B) Additionally, the mean overall hydrophobicity (determined based on Fauchere and Pliska, 1983) and the mean net charge are given. Note that the mean hydrophobicity of the FG repeats and the globular proteins does not significantly differ. This is remarkable given that FG repeats are intrinsically unfolded and hence expose their hydrophobic patches, whereas folded protein 'bury' their hydrophobic core (see main text for a detailed discussion). Unlike the protein of the PDB selection, FG repeats exclusively have a positive net charge. (C) Based on the parameters with the biggest variance (i.e. %Gly, %Thr and %NQ), the Nup98-derived FG repeats identified in the present study were plotted for visualization. The size of the bubbles reflects the hydrophobicity of the repeats. (D) To identify amino acids and parameters that tend to occur (or not to occur) together, the correlation of all parameters in the Nup98-derived FG repeat database was calculated. The derived Pearson's correlation coefficients were interpreted as effect sizes according to (Cohen, 1988). Note that there is a strong negative correlation between the overall hydrophobicity and NQ-content of the Nup98-derived FG repeats.

Yet, how do the Nup98-derived FG repeats vary from each other? Within the FG repeat database, the biggest variance was seen in the parameters glycine (%Gly), threonine (%Thr), asparagine and glutamine (%NQ) content. To visualize the dataset, all entries were plotted based on these three parameters in a bubble chart, with the size of the bubbles additionally reflecting the overall hydrophobicity of an entry (Figure 2.2.C).

Notably, most FG repeats are part of a large data cloud, with individual outliers that are either particularly rich in threonine (e.g. the GLFG repeats of the protozoan parasite *Theileria*, which causes sickness in humans and cattle), glycine (e.g. the GLFG repeats of *Perkinsus marinus*, a protozoan parasite of oysters) or asparagine/glutamine (e.g. the micronuclear *Tetrahymena thermophila* repeats). Remarkably, many of the outliers belong to the clade Chromalveolata, which combines a plethora of diverse species living in sundry environments with highly sophisticated live styles (such as multicellular organisms like kelp or parasitic/free-living unicellular protozoa). Arguably the most prominent examples are ciliates such as *Tetrahymena thermophila*, which contains two morphologically and functionally distinct nuclei in a common cytoplasm and harbors the most NQ-rich FG repeats known to date. The role of the latter in nuclear dualism will be subject of study in Section 2.4.

Taken together, two interesting questions arise: do individual parameters describing the Nup98-derived FG repeats correlate? And can different subgroups of FG repeats be defined based on their variance in the parameters used to describe them?

In an attempt to answer the first question, the pairwise correlation between each of the 27 parameters 'frequency of the twenty standard amino acids', 'NQ- and FILV-content', 'overall hydrophobicity', 'net charge' and 'FG motif occurrence' was calculated. Shown in Figure 2.2.D is a heat map graphically summarizing the results. The computed Pearson's correlation coefficients were interpreted as effect sizes, with absolute values ranging form 0.6–1.0 being considered indicative of a strong effect, absolute values in between 0.2–0.6 of a medium effect and values in the range of –0.2–0.2 of no effect, as suggested in the literature (Cohen, 1988).

As already suggested by the bubble plot (Figure 2.2.C), a very strong negative correlation between the NQ-content and the overall hydrophobicity of the FG repeats was indeed apparent (see also the plot below the heat map in Figure 2.2.D): the more hydrophobic a repeat is, the less likely it is to contain a high degree of NQ, and vice versa. Amongst the most hydrophobic FG repeats are the Nup98 homologs from plants and vertebrates, whereas in fungi, FG repeats are generally much more NQ-rich. Interestingly, the more hydrophobic FG repeats also contain more proline residues, which is especially the case for plants FG repeats.

Furthermore, the net charge of the FG repeats strongly (positively) correlates with their lysine content. Given that FG repeats generally are positively charged, contain only few charged residues and virtually lack arginine residues, this finding is not surprising. In FG repeats with higher charge density however, the negatively charged residues aspartate and glutamate strongly tend to occur together.

Strikingly, there is also a clear tendency towards a higher prevalence of more 'atypical' hydrophobic residues for charge-dense FG repeats (Figure 2.2.D). Yet, there is no correlation between the overall hydrophobicity and net charge. It thus seems plausible that the hydrophobic amino acids by and large might cover up the surplus of charged and hence extremely hydrophilic amino acids.

Finally, FG repeats that are rich in asparagine contain significantly fewer FG motifs and have the tendency to feature IF motifs instead. Also, such repeats by trend are less abundant in glycine and alanine residues. Prominent examples for these relations are the asparagine-rich IF repeats from *Tetrahymena thermophila* (see Section 2.4).

2.1.3 Cluster analysis of the FG repeat database.

Given the finding that such clear correlations and trends are detectable within the FG repeat database, it seems likely that based on their biochemical composition and biophysical properties, some FG repeats are more closely related than others, thus representing a defined subgroup ('cluster').

Recognition of such patterns in large datasets is the classical problem of the emerging 'knowledge discovery in databases' (KDD) field, which is not only of considerable interest in the biological sciences (Fayyad et al., 1996). Commonly, a variety of different clustering algorithms are employed to group individual entries of a large dataset into different clusters in an automated manner. These algorithms vary in multiple regards. For example, hierarchical clustering methods can be divided into divisive or agglomerative algorithms, depending on whether they part a complete dataset incrementally into smaller clusters, or merge multiple small clusters successively into larger ones, respectively. Moreover, algorithms that partition datasets into a manually fixed number of clusters are available (see e.g. Kaufman, 2005 for more details). Importantly, in contrast to most manual clustering approaches, the data is not curated and interpreted during automated clustering.

Hence, in order to be able to compare different clustering algorithms and interpret the results, a pre-defined control group comprised of globular proteins and the highly regular FSFG repeat region from the yeast FG Nup Nsp1 (which is extremely rich in charged amino acids; see also Section 2.2) was added to the database for the cluster analysis of the Nup98-derived FG repeats (see also Methods). Only clusterings in which the control group was faithfully identified and segregated were further analyzed and interpreted. In the end, nine different clusters were consistently found using varying clustering approaches, of which one was always the control group, as required (see Figure 2.3).

Generally, the clusters varied in the predominant FG motifs, the overall hydrophobicity and amino acid bias of the included FG repeats. The most divergent of all identified repeats belong to cluster 8 and are especially rich in asparagine, glutamine and isoleucine residues, whilst containing IF instead of FG motifs.

Table 2.1. *Cluster analysis* of the Nup98-derived FG repeat database.*

	Cluster 1	Cluster 2	Cluster 3	Cluster 4	Cluster 5	Cluster 6	Cluster 7	Cluster 8	Control Group
Predominant FG motif type	-	FG	FG	-	GLFG	FG	GLFG	IF	n.a.
Hydrophobicity	**2.0** *(+0.3)*	**2.1** *(+0.4)*	**2.2** *(+0.5)*	**1.55** *(-0.15)*	**1.3** *(-0.4)*	**1.6** *(-0.1)*	**1.35** *(-0.35)*	**1.1** *(-0.6)*	1.6
Amino acid bias	**Thr** (18.5%), **Gly** (11%)	**Ser** (19%), **Pro** (11%)	**Ala** (18%), **Pro** (9%)	**Ser** (13.5%)	**Asn** (17%), **Ser** (16%)	**Gly** (28.5%)	**Asn** (13%), **Gln** (12%)	**Asn** (23%), **Gln** (20%)	Charged/hydro-phobic aa
Species	*Homo sapiens* *Macacca mulatta* *Mus musculus* *Rattus norvegicus* *Canis familaris* *Bos taurus* *Pan troglodytes* *Gallus gallus* *Xenopus sp.* ***Branchiostoma floridae*** *Theileria annulata* *Naegleria gruberii* *Babesia bovis*	*Arabidopsis thaliana* *Daucus carota* *Oryza sativa* *Vitis vinifera* *Populus trichocarpa* *Ricinus communis* *Solanum lycopersicum* *Sorghum bicolor* *Zea mays* *Micromonas pusilla*	***Drosophila sp.*** *Anopheles sp.* *Ectocarpus siliculosus* *Thalassiosira pseudonana*	***Caenorhabditis sp.*** *Trichoplax adhaerens* *Tribolium castaneum* *Aspergillus sp.* *Encephalitozoon cuniculli* ***Giardia lamblia***	***Saccharomyces sp.*** *Schizosaccharo-myces sp.* *Candida sp.* *Pichia sp.* *Entamoeba sp.*	*Dictyostelium sp.* *Leishmania sp.* ***Trypanosoma sp.*** *Toxoplasma gondii* *Perkinsus marinus* *Bigelowiella natans* *Aspergillus sp.* *Ajellomyces sp.* *Candida sp.* *Penicillium sp.*	*Tetrahymena thermophila* *Paramecium tetraurelia* *Saccharomyces sp.* *Candida sp.* *Ashyba sp.* *Entamoeba sp.*	*Tetrahymena thermophila* *Paramecium tetraurelia*	**Nsp1 FxFG repeats** GFP MBP mCherry Myosin *E. coli* metabolic enzymes

*To analyze the (compositional) relationship between the identified FG repeats, an unbiased cluster analysis was performed (see Methods for details). For comparison of different clustering algorithms, a control group comprising (i) the charge-rich Nsp1 FxFG repeats derived from the $Nsp1_{274-601}$ domain, (ii) the inert permeation probes commonly used in this study (i.e. GFP, mCherry, MBP) and (iii) a range of metabolic enzymes from *E. coli* was included. Only such clusterings in which the control group was faithfully separated from the FG repeat-containing clusters were accepted. Ultimately, the nine shown clusters were consistently identified with different clustering algorithms. **Legend:** indicated are (i) the predominant FG motif, (ii) the hydrophobicity (the numerical values represent the mean of the clusters and the difference to total mean of the complete database) and (iii) the most obvious amino acid bias of the FG repeats in a given cluster (cluster means given). The abbreviation *sp.* represents multiple species of a given genus. The FG repeats of the species highlighted in bold have been further analyzed in the present study. The color code represents the classification of each species within the eukaryotic tree, as introduced in Figure 2.1.

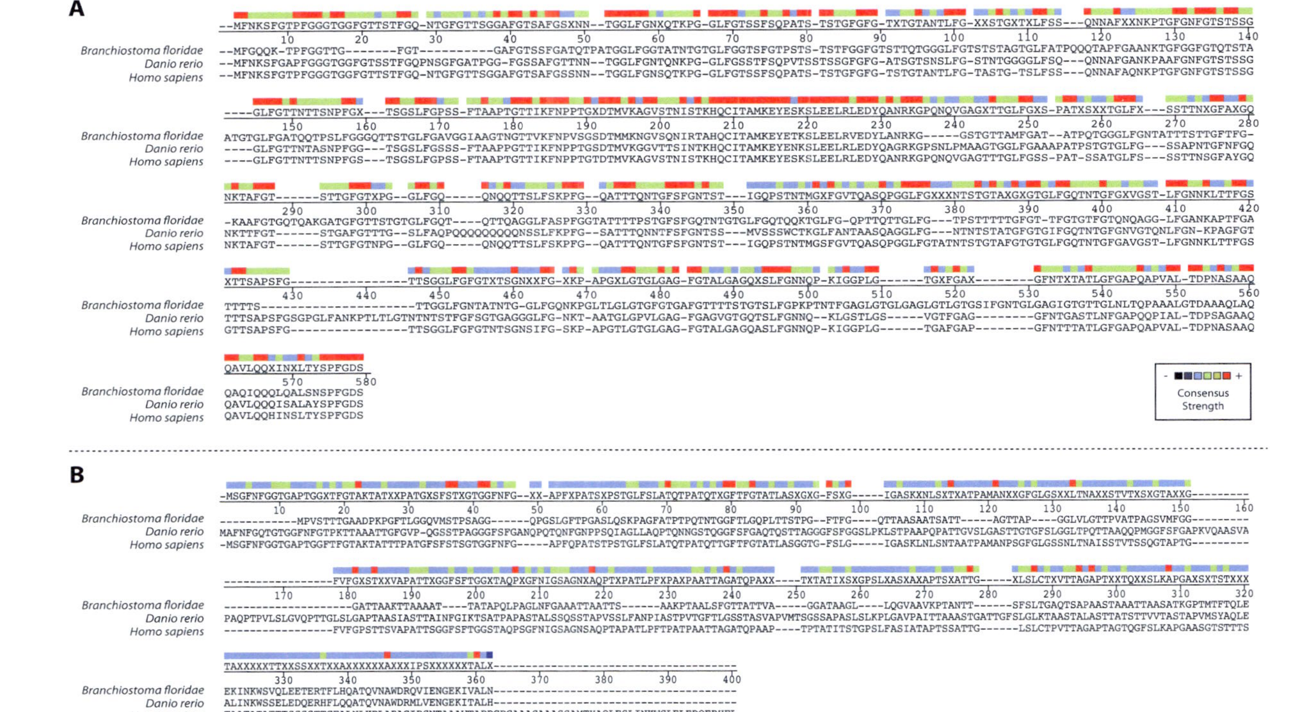

Figure 2.3. Amino acid sequence alignments of the FG repeat domains from (A) Nup98 and (B) Nup62. The Nup98 and Nup62 homologs of *Branchiostoma floridae*, *Danio rerio* and *Homo sapiens* were identified via their C-terminal anchoring domains (i.e. the Nucleoporin2 autopeptidase domain of Nup98 or the heptad repeat coiled-coil region of Nup62). The ClustalW algorithm was used to align the FG repeat regions (boundaries according to A. Labohka, personal communication). Note that the Nup98 GLFG repeats are surprisingly conserved from humans to zebrafish, which is not the case for the Nup62 FG repeats. This finding likely reflects the crucial role Nup98 plays in the context of the NPC permeability barrier (see Hülsmann et al., 2012 and main text).

Notably, only a subset of all identified Nup98 homologs within the closely related Chromalveolates *Tetrahymena* and *Paramecium* fall into this group, namely their micronuclear Nup98 orthologs (see Section 2.4 for details).

Three further noteworthy observations were made. First, especially the FG repeats of the earliest branching eukaryotes (i.e. the Excavates and Euglenozoa), as well as the fungal Nup98 homologs, scatter across multiple diverse clusters. Second, Cluster 2 was a true monophyletic group, encompassing exclusively FG repeats from the kingdom Plantae (i.e. plants, green and red algae). Third, when comparing the sequences in cluster 1, it was found that the Nup98 FG repeats of chordates, reaching almost back to the last common ancestor between lancelets (*Branchiostoma floridae*) and zebrafish (*Danio rerio*), surprisingly are nearly identical in sequence and spacing of their GLFG motifs (see Figure 2.3.A).

Based on the presented cluster analysis, representative FG repeats from all clusters were chosen for further detailed analysis, covering each eukaryotic clade as well as the observed repeat diversity (see Table 2.2. for the amino acid sequences of selected representative FG repeats from each cluster). Precisely, the FG repeats of the Nup98 orthologs from (i) *Homo sapiens, Drosophila melongaster, Caenorhabditis elegans, Branchiostoma floridae* and *Saccharomyces cerevisiae* (all **Opisthokonta**), (ii) *Dictyostelium discoideum* (**Amoebozoa**), (iii) *Arabidopsis thaliana* and *Mircomonas pusilla* (**Plantae**), (iv) *Tetrahymena thermophila* (**Chromalveolata**), (v) *Bigelowiella natans* (**Rhizaria**), (vi) *Giardia lamblia* (**Excavata**) and (vii) *Trypanosoma brucei* (**Euglenozoa**) were selected to challenge the hypothesis that the cohesiveness of FG repeats is evolutionary conserved in all Eukaryotes.

Table 2.2. *Amino acid sequence of representative FG repeats for each Cluster.*

Cluster 1 – *Homo sapiens*

```
MFNKSFGTPFGGGTGGFGTTSTFGQNTGFGTTSGGAFGTSAFGSSNNTGGLFGNSQTKPGGLFGTSSFSQPATSTSTGFGF
GTSTGTANTLFGTASTGTSLFSSQNNAFAQNKPTGFGNFGTSTSSGGLFGTTNTTSNPFGSTSGSLFGPSSFTAAPTGTTI
KFNPPTGTDTMVKAGVSTNISTKHQCITAMKEYESKSLEELRLEDYQANRKGPQNQVGAGTTTGLFGSSPATSSATGLFSS
STTNSGFAYGQNKTAFGTSTTGFGTNPGGLFGQQNQQTTSLFSKPFGQATTTQNTGFSFGNTSTIGQPSTNTMGSFGVTQA
SQPGGLFGTATNTSTGTAFGTGTGLFGQTNTGFGAVGSTLFGNNKLTTFGSGTTSAPSFGTTSGGLFGFGTNTSGNSIFGS
KPAPGTLGTGLGAGFGTALGAGQASLFGNNQPKIGGPLGTGAFGAPGFNTTTATLGFGAPQAPVALTDPNASAAQQAVLQQ
HINSLTYSPFGDS
```

Cluster 2 – *Arabidopsis thaliana*[1]

```
MFGSSNPFGQSSGTSPFGSQSLFGQTSNTSSNNPFAPATPFGTSAPFAAQSGSSIFGSTSTGVFGAPQTSSPFASTPTFGA
SSSPAFGNSTPAFGASPASSPFGGSSGFGQKPLGFSTPQSNPFGNSTQQSQPAFGNTSFGSSTPFGATNTPAFGAPSTPSF
GATSTPSFGASSTPAFGATNTPAFGASNSPSFGATNTPAFGASPTPAFGSTGTTFGNTGFGSGGAFGASNTPAFGASGTPA
FGASGTPAFGASSTPAFGASSTPAFGASSTPAFGGSSTPSFGASNTSSFSFGSSPAFGQSTSAFGSSAFGSTPSPFGGAQA
STPTFGGSGFGQSTFGGQQGGSRAVPYAPTVEADTGTGTQPAGKLESISAMPAYKEKNYEELRWEDYQRGDKGGPLPAGQS
PGNAGFGISPSQPNPFSPSPAFGQTSANPTNPFSSSTSTNPFAPQTPTIASSSFGTATSNFGSSPFGVTSSSNLFGSGSST
TTSVFGSSSAFGTTTPSPLFGSSSTPGFGSSSSIFGSAPGQGATPAFGNSQPSTLFNSTPSTGQTGSAFGQTGSAFGQFGQ
SSAPAFGQNSIFNKPSTGFGNMFSSSSTLTTSSSSPFGQTMPAGVTPFQSAQPGQASNGFGFNNFGQNQAANTTGNAGGLG
IFGQGNFGQSPAPLNSVVLQPVAVTNPFGTLPA
```

Cluster 3 – *Drosophila melongaster*

```
MFGGAKPSFGATPAATSFGGFSGTTTTTPFGQSAFGKPAAPAFGNTSTFAAQPAQQSLFGAAATPAQPAGGLFGANTSTGF
GSTATAQPTAFGAFSQPQQTSNIFGSTQTAASTSLFGQSTLPAFGAAKPTMTAFGQTAAAQPTGSLFGQPAAATSTTGFGG
FGTSAPTTTNVFGSGTASAFAQPQATAVGASGVNTGTAVAKYQPTIGTDTLMKSGQANSVNTKQHCITAMKEFEGKSLEEL
RLEDYMCGRKGPQAGNAPGAFGFGAQVTQPAQPASGGLFGSTAQPSTGLFGQTVTENKSMFGTTAFGQQPATNNAFGAATQ
QNNFLQKPFGATTTTPFAAPAADASNPFGAKPAFGQGGSLFGQAPATSAAPAFGQTNTGFGGFGTTAGATQQSTLFGATPA
ADPNKSAFGLGTAASAATTGFGFGAPATSTAGGGLFGNKPATSFAAPTFGATSTASTPFSNFGLNTSTAATGGGLFNSGLN
KPATSGFGGFGATSAAPLNFNAGNTGGSLFGNTAKPGGGLFGGGTTTLGGTGAAPTGGLFGGGTTSFGGVGGSLGGGGFGM
GTNNSLTGGIMGA
```

Cluster 4 – *Caenorhabditis elegans*

```
MFGQNKSFGSSSFGGGSSGSGLFGQNNQNNQNKGLFGQPANNSGTTGLFGAAQNKPAGSIFGAASNTSSIFGSPQQPQNNQ
SSLFGGGQNNANRSIFGSTSSAAPASSSLFGNNANNTGTSSIFGSNNNAPSGGGLFGASTVSGTTVKFEPPISSDTMMRNG
TTQTISTKHMCISAMSKYDGKSIEELRVEDYIANRKAPGTGTTSTGGGLFGASNTTNQAGSSGLFGSSNAQQKTSLFGGAS
TSSPFGGNTSTANTGSSLFGNNNANTSAASGSLFGAKPAGSSLFGSTATTGASTFGQTTGSSLFGNQQPQTNTGGSLFGNT
QNQNQSGSLFGNTGTTGTGLFGQAQQQPQQQSSGFSFGGAPAATNAFGQPAAANTGGSLFGNTSTANTGSSLFGAKPATST
GFTFGATQPTTTNAFGSTNTGGGLFGNNAAKPGGLFGNTTNTGTGGGLFGSQPQASSGGLFGSNTQATQPLNTGFGNLAQP
QIVMQQQ
```

Cluster 5 – *Saccharomyces cerevisae*[2]

```
FTFGNQNTSTPTSTPAQPSSSLQFPQKSTGLFGNVNVNANTSTPSPSGGLFNANSNANSISQQPANNSLFGNKPAQPSGGL
FGATNNTTSKSAGSLFGNNNATANSTGSTGLFSGSNNIASSTQNGGLFGNSNNNNITSTTQNGGLFGKPTTTPAGAGGLFG
NSSSTNSTTGLFGSNNTQSSTGIFGQKPGASTTGGLFG
```

Cluster 6 – *Dictyostelium discoideum*

```
MFGGQFGSFGAKPAATASPFGAPSAAPTTSLFGSTAPSSGFGGFGSTAQTTQPTTGGFGGFGGFGGATTTQQPAASPFGGG
GTGGSGLFGSSAQTTTQQPGASPFGGGFGTTTTTTTQQPGASPFGGTGGGLFGSSAQTTTQQQGASPFGGFGGATTTQPSL
LSGATGGFGGFGGSTTSGTQLGGGGGATSGAFGGSSSPFGGSGGATTSSPFGGGGGSFGATTQKQYGTPIPYQQTTIEGNT
FVSISAMPQYNDRSFEELRFEDITHRKDIVYKTGGGSGGGNSLFGSTPTTQPSSPFGAQTTTQTTGGLFGGQTTTSPFGGQ
TSATPGSSLFGSTQPTQQQTSGGLFGSVQPTQQQAGGGLFGSMPSTGGSSLFGSTQPTQQQTGGAQPTQSLFGGQTQTTTS
PFGSQTSTPFGQPQQTNTGSGLFGAQQTQQTNTGGGLFGAQPTQQTSGGGLFGTQPTSGTGLFGTSPTAGGTGLFGTTQPT
SQGTGLFGTTQPTTQGTGLFGTSPTSGTGLFGSTPTSGTGLFGSTPTSGTGLFGSAQPPQNQQSQTSLFGNTGTGTGATNTGT
GLFGSAQPSSNPGGGLFGSAQPSTTTGGLFGSNQPTAQPTTSLFGNTTGSVGGLGATPNITSGLFGSNPAQTGGLFGSTQP
TTQTSLFGNTGSTGGLGAQNGGGLFGNLSQPTATAGQGLSGGLFGNLSQPTATAGQGLSSGGLFGNTLLG
```

Cluster 7 – *Tetrahymena thermophila* (MAC)[3]

```
MFGNTGGGGLFGNTQTQQTGGGLFGQPQQTQFGQTGATGGGLFGGATNTFGGGGGGGLFGGNNNQQTNPTAGGGIFGQGTT
GLGGAPAQTGGGLFGAPQNNQGGGLFGGGTTTGGGMFGNQANTQTGGGGLFGGPSQPTTQPPAFSLNNPTTGGGGLFGQPA
NTMGGNNGGLFGGQTNSFGANNNMLGNNNRPQGAGIFGGATTTAPTGNTGMFGGIGANNGGGGLFGMNNTNTNPTGGFGAT
NPTAGGGGLFGGGATTTGGGGLFGGGNTQGGGLLGTANTTAGGLLGGGFNMNNNTGGILGQTNNQFGLGSFGTNNNAAAAP
FQPKASANGVLTKPNEKNLCYAISNGTDFCIFELALTQRKLVKAGQLKPGAQQAGGMFGQPAQGGNGLFGGGGAATTTPFG
GAQNGNLFGGQNTQAQGGGLFGAPVNNAATGAGGGLFGAKPAATTTGGGLFGQMPAQTGGFLGNTATQPAGGGLFGGATTT
QAPGGGGGGGLFGGNTTAATTGGGLFGGNTQTGGATGGLFGGQQPNNQGGLFLNTGNANNANTGGGLFGGATTTPATGGGL
FGGSTNTQPGLATGGGLFGNNQGASQPAAQGGLFGGAAPQQNSLFGGATAGGQTGGLFGGATGATQQQQGGGLFGQTASNPT
QGGGLFGAANPGLGGAAA
```

Cluster 8 – *Tetrahymena thermophila* (MIC)[4]

```
MYGYQNVYNPYMSNSQQQYGYPQNMNPYNNPQYMTEQQYQNQNQMNYLANGVYNNNMNRSVQQIPNQNIFQSNNQNVMNSN
PINPIVASRMNRQTSNPVGGSGNQNNQMNRNMYASYNNANPIFNQNQVNNNNMYYNNSNQTANFGQNTQGYYNNQNINNQS
IFNGNNNTQRNFNNQNQSIFNNNNNNTNNNLNIFSNSNNNQNVFNNTNNSQNIFKNNNNNNNQSNFNNNNSSIFNNNNNGN
NQNIFSKNNNNNQSIFNNNNNNNNNIFKSSNNNQNIFSSNNNNNNNSSIFNNTNQSILNNNSSIFNNSNNQSIFNNSMNN
NLNVNRSNNQSIFNNNNNNTIQNVFNQPNSILNQPNQTQSIFGNAPN
```

Legend: Highlighted in blue are different flavors of FG motifs, in green glycine, in purple threonine and serine residues, in red charged residues and in grey asparagine and glutamine residues. **Remarks:** FG domains of [1] AtNup98A (see also Tamura et al., 2010), [2] ScNup145 (K. Kirli, personal communication), [3] TtMacNup98A and [4] TtMicNup98A (for both see also Iwamoto et al., 2009).

2.1.4 Cloning, expression and purification of the Nup98-derived FG repeats and their ability to form hydrogels.

Given the number of different model organisms and the fact that ciliates such as *Tetrahymena* deviate from the universal genetic code[1], synthetic genes optimized for the recombinant bacterial expression of the aforementioned repeats were designed, commercially synthesized and sub-cloned into appropriate expression vectors (see Materials).

Note that due to their extreme AT-content (i.e. more than 60%), the genes encoding for the asparagine-rich *Tetrahymena* repeats were extremely unstable in the bacterial cloning and expression host *Escherichia coli*. To solve this problem, a single-copy expression vector based on the bacterial F-plasmid was created and the *E. coli* SURE2 strain used, which lacks a number of crucial components required for homologous recombination (*rec*B, *rec*J), palindrome excision (*sbc*C) and nucleotide excision repair (*uvr*C).

Whereas for most FG repeats, the expression levels were fine, the *Tetrahymena* and *Bigelowiella* proteins only expressed poorly in *E. coli* even after optimization of the expression conditions. The use of specially designed custom His-tags helped to slightly improve the expression levels and facilitate the purification procedure. The tags were tailored to meet a number of criteria, both on the protein and nucleotide level. First, the amino acid sequence comprised mainly pairs of histidines interspersed with glycines to bestow a high flexibility to the tags. Second, the nucleotide sequences encoding the tags were optimized to lack RNA secondary structure around the ribosome binding site plus the first 30 codons once transcribed, which reportedly is crucial for efficient translation (see e.g. Kudla et al., 2009; Dokyun Na, 2010). Third, to further enhance expression, the coding sequences were designed such that a 5' codon composition bias towards AT-rich codons was introduced (see e.g. Brock et al., 2007).

Ultimately, all FG repeats could be purified under denaturing conditions using Ni^{2+}-affinity chromatography and reverse-phase HPLC. Afterwards, the proteins were lyophilized (see Methods).

Hydrogels were formed from the lyophilized proteins by re-suspending appropriate amounts (gravimetrically determined) in 0.2% TFA at a standard concentration of 200mg/ml (unless stated otherwise). After gelation, the gels were always neutralized by equilibration with buffer (see Methods).

These initially extreme conditions were chosen to allow for the formation of a sufficiently concentrated and homogenous starting solution prior to the onset of gelation

---

[1] Interestingly, the universal stop codons UAA and UAG are translated into glutamine in most ciliates, including *Tetrahymena* (Tourancheau et al., 1995). In fact, they are the preferred codons for this amino acid. Hence, working with *Tetrahymena* cDNA in bacteria was not feasible, primarily because of two reasons: first, the two additional glutamine codons make *Tetrahymena* DNA extremely AT-rich, which is generally a problem for *E.coli* (see e.g. Godiska et al., 2010). Second, too many site-directed mutations would have been required to make the genetic codes of ciliates and bacteria compatible again. The solutions found to circumvent these problems are described in the main text.

(note that in some cases, even small concentrations of the chaotropic reagent guanidine hydrochloride had to be additionally added). Indeed, such means of controlling and timing hydrogel formation are essential for many applications, ranging e.g. from casting agarose gels in the lab to preparing gelatin dishes in the kitchen. Whereas in the case of FG repeats, gelation is accompanied by a pH shift, both agarose and gelatin gels are however formed in response to a temperature shift.

A general observation during hydrogel formation was that the greater the overall hydrophobicity of the FG repeats, the stickier and harder to re-suspend were the lyophilized proteins. This was especially problematic for the *Arabidopsis thaliana*, *Homo sapiens* and *Branchiostoma floridae* repeats, which have a great tendency to aggregate instantaneously upon solvent addition. The aforementioned custom HisTags significantly helped to improve the situation, as the imidazole side chains contributed extra charge at low pH, thus facilitating the generation of homogenous concentrated protein solutions. In contrast, the NQ-rich repeats from *Saccharomyces* and *Tetrahymena*, as well as the only slightly hydrophobic *Caenorhabditis* repeats, were especially well behaved during hydrogel formation.

In summary, following the approaches outlined above, all tested Nup98-derived FG repeats were found to be cohesive and able to form hydrogels.

2.1.5 The size of Ran is conserved in eukaryotes and might define the minimal fineness of the NPC size-sieve.

Although the nucleus and cytoplasm freely exchange ions and small metabolites, their protein composition is fundamentally different. The key task of the NPC permeability barrier is to prevent the complete intermixing of the nuclear and cytoplasmic compartments, whilst allowing the selective transport of all required factors to equip the two specialized 'reaction chambers' with their own characteristic subset of molecular players. The barrier operates in a size-dependent manner: molecules above a certain size can no longer efficiently traverse the NPC, unless they are chaperoned by NTRs (see also Section 1.3).

Thus, if the NPC permeability barrier were indeed made of a hydrogel-like selective phase, the hydrogels formed by the Nup98-derived FG repeats should on the one hand exclude large molecules and on the other hand allow the entry of NTR•cargo complexes, as it has already been shown for Nsp1 (Frey and Görlich, 2007), Nup49 and Nup57 (Frey and Görlich, 2009), and *Xenopus* FG repeat (A. Labohka, personal communication and Milles and Lemke, 2011) hydrogels.

Yet, exactly how selective must the NPC permeability barrier and hence FG repeat hydrogels be? Active importin- and exportin- mediated nucleocytoplasmic exchange is fueled by the RanGTP gradient across the nuclear envelope (see Section 1.1). Ran is a small GTPase with a diameter of approximately 4.5nm in humans (as determined from the

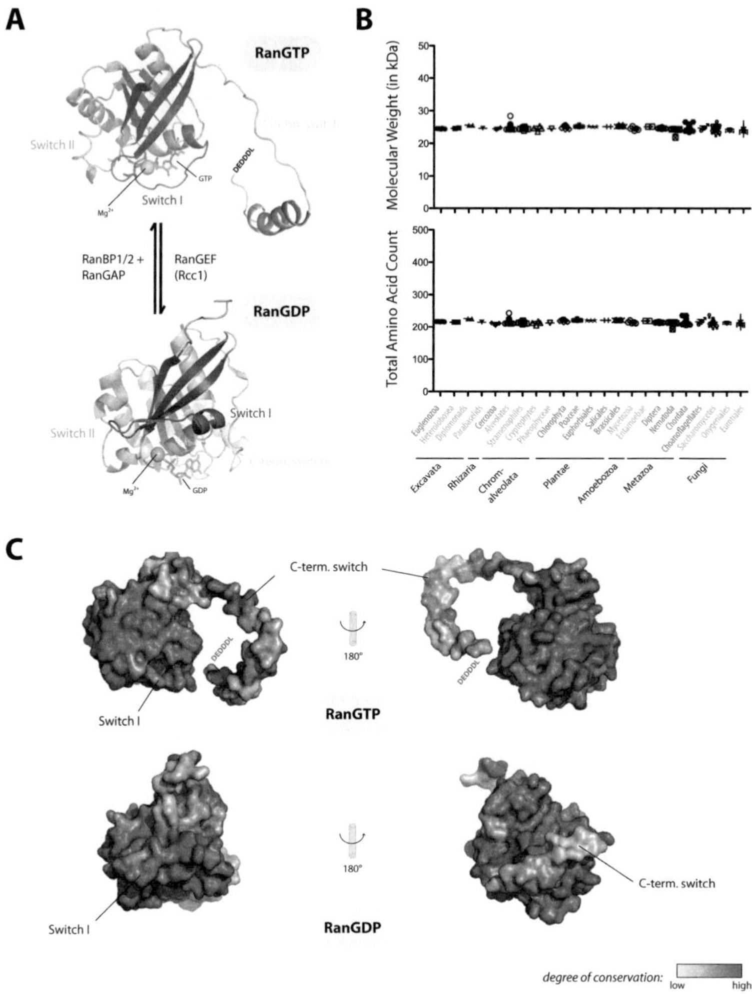

Figure 2.4. The size of Ran is evolutionary highly conserved in eukaryotes. The GTPase Ran is a molecular switch controlling NTR function. Shown here are the structures of Ran in its GTP- (Vetter et al., 1999) and GDP-bound (Partridge and Schwartz, 2009) form (A). The very C-terminal residues of the molecule were not resolved in the electron density map and are hence only schematically depicted here. Note the drastic conformational changes the molecule undergoes upon GTP hydrolysis (which requires the help of RanBP1/2 and RanGAP) or GTP loading (aided by Rcc1), especially in the three highlighted switch regions. The role of RanGTP and its interaction with NTRs are discussed in the main text. In short, a step RanGTP gradient across the nuclear envelope drives nucleocytoplasmic transport. Hence, the NPC permeability barrier must prevent the passive leakage of RanGTP out of the nucleus and consequently sieve-out molecules in the size range of Ran. The size of Ran is evolutionary highly conserved (B) and thus marks the universally valid critical sieving size. The degree of amino acid conservation was mapped onto the Ran structures (C).

crystal structure published by Partridge and Schwartz, 2009; PDB entry 3GJ0), which undergoes significant conformational changes depending on its nucleotide state (see Figure 2.4.A and references therein). Notably, breakdown of the RanGTP gradient would cause nucleocytoplasmic transport to cease, eventually leading to life-threatening cellular havoc (Izaurralde et al., 1997; Görlich et al., 2003). Hence, the human NPC permeability barrier must already efficiently restrict the passive diffusion of objects larger than ~4-5nm in diameter to prevent the collapse of the RanGTP gradient.

In agreement with this conclusion, a recent thorough analysis of the passive diffusion of differently-sized inert molecules across intact NPCs in a permeabilised HeLa cell system suggested that the dominating radius of the passive NPC diffusion channel is ~2.6nm (Mohr et al., 2009).

To establish a universal size cut-off, homologs of Ran were systematically identified in all eukaryotic clades following the same approach used for the discovery of FG repeats (see Methods).

Strikingly, the size (in kDa) and even the sequence length (in amino acids) of Ran is extremely well conserved (see Figure 2.4.B). Statistically significant differences (i.e. $P<0.005$ in a Kruskal-Wallis test) are only observed for the plant lineages *Poacea* and *Brassicales*. Here, the Ran homologs are altogether slightly above 25kDa, whereas on average the molecular weight of Ran is slightly below 25kDa. Nevertheless, these are negligible differences in regard to the expected size selectivity and thus it is safe to assume that also the dimensions of Ran are highly conserved in all eukaryotes.

The most divergent Ran homologs identified, coming from the early-branching excavate *Giardia* and human, share more than 50% sequence identity. As Figure 2.4.C shows, the highest degree in sequence conservation generally applies to the β-strand core of Ran, as well as the switch I and switch II regions. In contrast, the amino acid composition and length of the C-terminal switch however varies, but importantly always contains a functionally relevant terminal acidic patch.

In summary, based on the conservation of Ran, the upper size limit for noticeable passive diffusion across NPCs was predicted not to exceed ~4nm. Hence, FG repeats hydrogels should have similar size-sieving capabilities if they were to form permeability barriers with NPC-like properties.

2.1.6 Most studied FG hydrogels were found to be molecular sieves as stringent as NPCs.

To explore this notion systematically, polyethylene glycol (PEG) molecules varying in their length were labeled with fluorescent dyes, separated according to their size (see Methods) and used as permeation probes to determine the sieving properties of selected FG repeat hydrogels (see below).

In short, the hydrogel permeation assay (Figure 2.5) works as follows: the lyophilized FG repeats of interest are re-suspended in an appropriate solvent and pipetted in small drops (>1.0 µl) onto microscopy slides (see Methods). After equilibration with assay buffer, the slides are mounted to a confocal microscope and aligned such that half the imaging window covers the hydrogel phase, whereas the other half records the buffer side. After addition of a fluorescently-labelled probe, its influx into the hydrogel is recorded at defined time intervals and quantified using custom-written Mathematica routines (see Methods). The enrichment of a given species in the gel is ultimately expressed relative to its fluorescence intensity in the buffer side.

In theory, the enrichment of inert molecules (such as the PEG species) in FG hydrogels should hence be well below a partitioning coefficient of 1.0 (indicative of equilibration between the buffer and gel phase), even after extended periods of incubation. Very good values are in the range of 0.2-0.3, as here, the expected volume fraction of the inert molecule in the gel is far below its actual volume fraction in the bulk solvent.

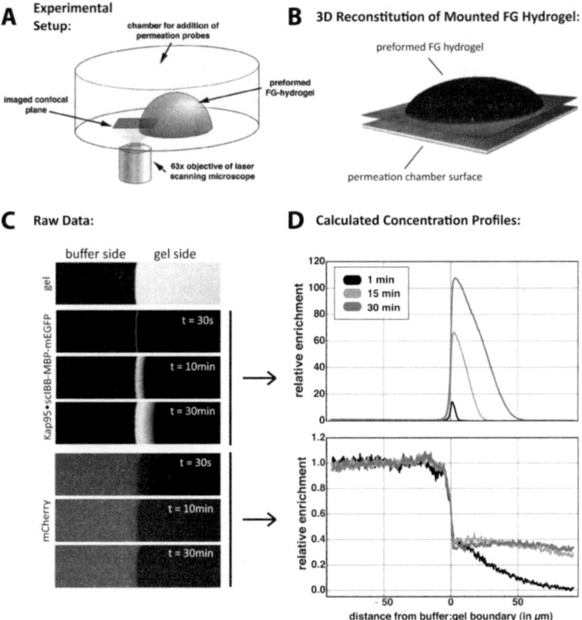

Figure 2.5. The hydrogel permeation assay setup. (A) Hydrogels are formed in special permeation chambers (µ-slides, Ibidi) and mounted to a laser scanning confocal microscope such that half of the imaging field covers the buffer or gel side, respectively (courtesy of S. Frey). (B) 3D-reconstitution of a mounted Nsp1 hydrogel prior to experimentation, computed based on a 3D-stack through the gel. The gel was detected via an Atto647N-labelled tracer molecule. The standard concentration of the hydrogels used in this study was 200mg/ml. (C) Exemplary, the raw data and the thereon-calculated concentration profiles (D) for a standard experiment with an active NTR•cargo species (here: Kap95•IBB-MPB-GFP; used at 1µM) and an inert permeation probe (here: mCherry; used at 3µM) penetrating into a Nsp1 hydrogel are shown. For

quantification, the concentration of either species in the buffer was normalized to one: the relative enrichment of a given species in the indicated hydrogel is shown throughout this study.

Hydrogels formed from Nup98-derived FG repeats of each cluster, namely the species *Homo sapiens* (Cluster 1), *Arabidopsis thaliana* (Cluster 2), *Drosophila melongaster* (Cluster 3), *Caenorhabditis elegans* (Cluster 4), *Saccharomyces cerevisiae* (Cluster 5), *Dictyostelium discoideum* (Cluster 6) and *Tetrahymena thermophila* (Cluster 7 and 8), were challenged with the PEG probes.

Surprisingly, whereas most hydrogels indeed posed firm barriers towards PEG molecules with radii of ≥2.5nm, the NQ-rich repeats from *Saccharomyces* (Cluster 5) and *Tetrahymena* (Cluster 8) formed hydrogels that did not show any sieving effect (Figure 2.6). In fact, with partitioning coefficient above 1.0 even for large species with r ≈ 4.15nm, most of the PEG probes even slightly enriched in these gels.

In contrast, the human, *Drosophila* and *Arabidopsis* repeats formed hydrogels that even impaired the entry of the smallest PEG species (Figure 2.6). After o/n incubation, the observed partitioning coefficients were only in the range of 0.5-0.6, which means that the in-gel volume fraction of the PEG is still well below its volume share in the buffer side. Notably, the radius of this species was ~1.4nm, which is well below the expected size limit of NPCs. The macronuclear *Tetrahymena*, *Caenorhabditis* and *Dictyostelium* repeats could however only efficiently restrict the passage of PEG molecules with a radius of up to 2.5nm.

Taken together, most studied FG repeats formed hydrogels with a size selectivity matching or outperforming the NPC . The notable exceptions were the NQ-rich FG repeats of *Saccharomyces* (Cluster 5) and *Tetrahymena* (Cluster 8).

2.1.7 FG hydrogels reconstitute facilitated diffusion of NTR•cargo complexes only to varying degrees.

Apart from restricting the free diffusion of molecules above the size limit, the NPC permeability barrier must also grant larger NTR•cargo complexes to pass. Consequently, this should also apply to FG repeat hydrogels if they were indeed the functional hallmark of NPCs. To test the active permeability properties of the Nup98-derived FG repeat hydrogels, they were challenged with a complex of human Impβ and IBB-MBP-GFP. As an internal control, an MBP-mCherry fusion was always applied in parallel.

Remarkably, the three most stringent sieves, i.e. the human, *Drosophila* and *Arabidopsis* FG hydrogels, hardly allowed the facilitated influx of the Impβ•IBB-MBP-GFP complex, which reached only partitioning coefficients of up to ~15-50 and diffusion depths of only up to ~10-15µm after 30 minutes of incubation (Figure 2.7). In contrast, the *Caenorhabditis* gel enriched this active species ~250-fold, which in the same timespan also diffused 30-

A

PEG-4 (1.4 nm)

PEG-3 (2.5 nm)

PEG-2 (3.4 nm)

PEG-1 (4.15 nm)

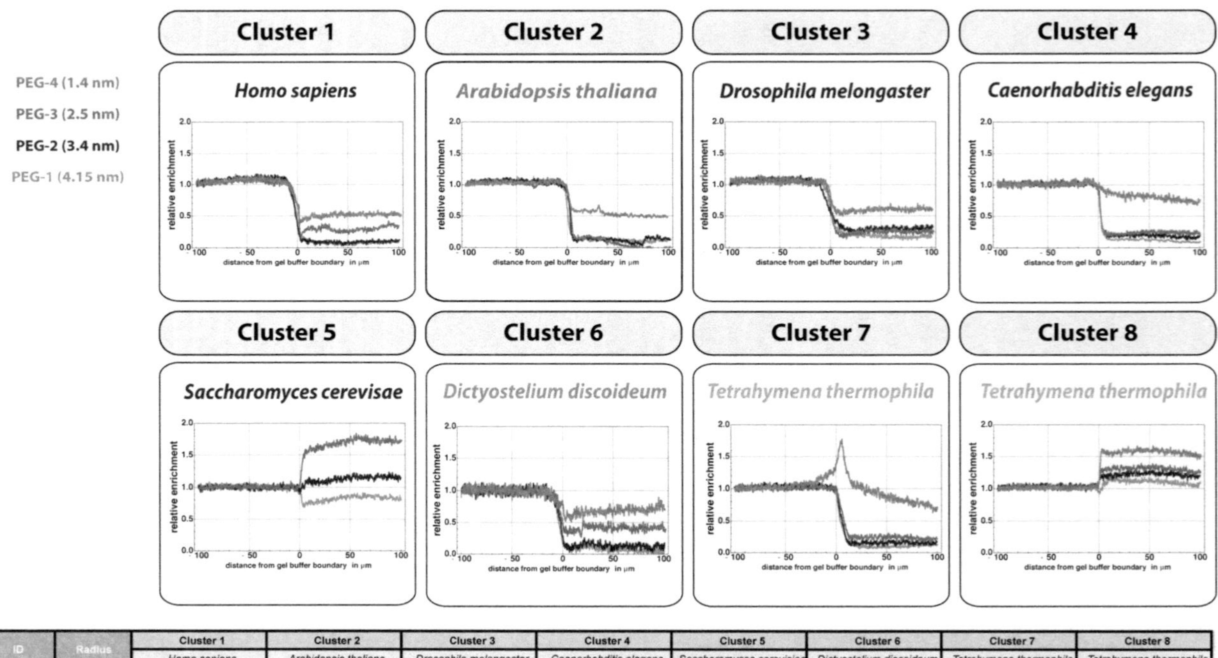

B

ID	Radius	Cluster 1	Cluster 2	Cluster 3	Cluster 4	Cluster 5	Cluster 6	Cluster 7	Cluster 8
		Homo sapiens	Arabidopsis thaliana	Drosophila melongaster	Caenorhabditis elegans	Saccharomyces cerevisiae	Dictyostelium discoideum	Tetrahymena thermophila	Tetrahymena thermophila
PEG-1	4.15 nm	0.1	0.15	>0.2	0.1	0.85	0.1	0.1	1.1
PEG-2	3.4 nm	0.1	0.15	0.3	0.2	1.15	0.15	0.15	1.2
PEG-3	2.5 nm	0.2	0.15	0.25	0.25	1.75	0.4	0.25	1.3
PEG-4	1.4 nm	0.5	0.55	0.6	0.85	>2	0.7	1.0-1.3	1.6

Figure 2.6. Sieving capacity of the Nup98-derived FG hydrogels. To systematically test whether the hydrogel barriers formed by FG repeats can in general match the size-sieving capabilities of the NPC, hydrogels formed from the purified FG domains of the Nup98 homologs from the indicated species were challenged o/n at RT with differently sized (i.e. r=1.4, 2.5, 3.4 or 4.15 nm), Atto488-labelled PEG species. Shown are the concentration profiles (A) and a tabular summary of the final partitioning coefficients (B). Note that in contrast to all others, the hydrogels derived from the NQ-rich Saccharomyces and Tetrahymena FG repeats cannot be considered selective NPC-like barriers.

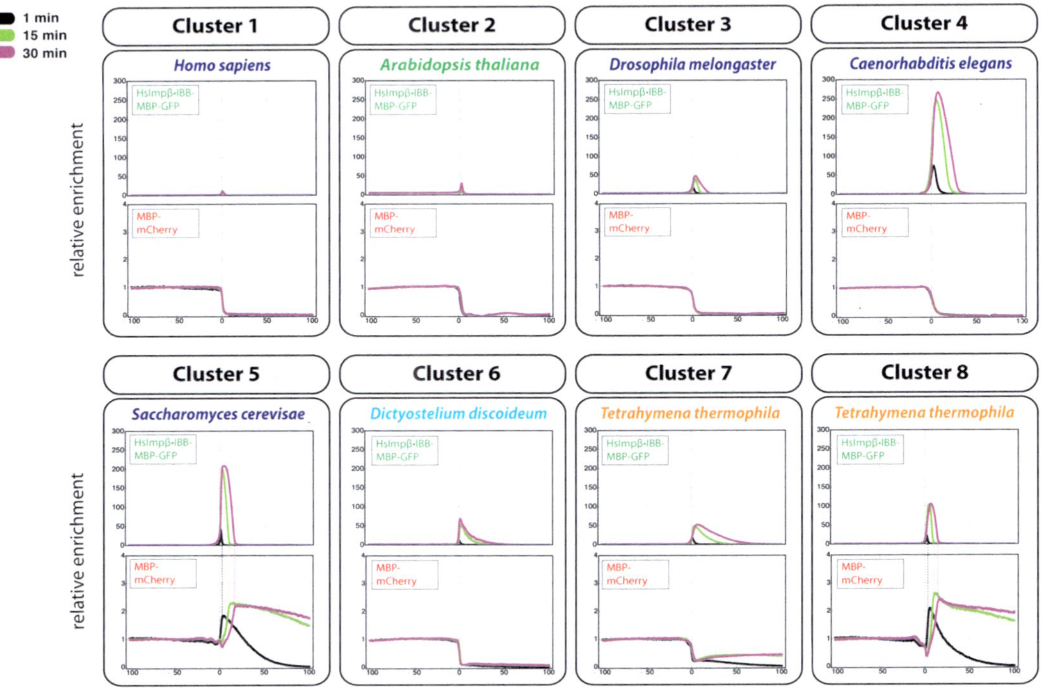

Figure 2.7. Most FG repeat hydrogels reproduce the fundamental properties of the NPC permeability barrier. The indicated hydrogels were each simultaneously challenged with a preformed complex of human Impβ and IBB-MBP-GFP (at 1µM), as well as the passive permeation probe MBP-mCherry (at 3µM). Influx of both species was followed for 30 minutes in 60-second intervals. Shown are the concentration profiles for t=1, 15 and 30 minutes. Note that the FG repeats from cluster 4 and 6 formed hydrogels that show a significant background binding of the MBP-mCherry species. However, in the parts of the gels where the NTR•cargo complex enriched, the passive permeation probe was displaced. See main text for further discussion of the results.

40µm into the gel. As expected, all four hydrogels also efficiently excluded the inert MBP-mCherry control.

The latter was strikingly not true for the hydrogels formed from the NQ-rich *Saccharomyces* and *Tetrahymena* FG repeats, which enriched the MBP-mCherry species ~2-fold. Yet, they allowed the uptake and intra-gel diffusion of Impβ•IBB-MBP-GFP.

Finally, the canonical *Tetrahymena* and the *Dictyostelium* FG repeats formed hydrogels that were found to be good passive barriers and allowed the enrichment and intra-gel diffusion of NTR•cargo complexes to a similar extend as published before for Nsp1-derived FG/FxFG hydrogels (Frey and Görlich, 2009).

In summary, the Nup98-derived FG repeats do not only vary in their amino acid composition, but also form hydrogels with different permeability properties.

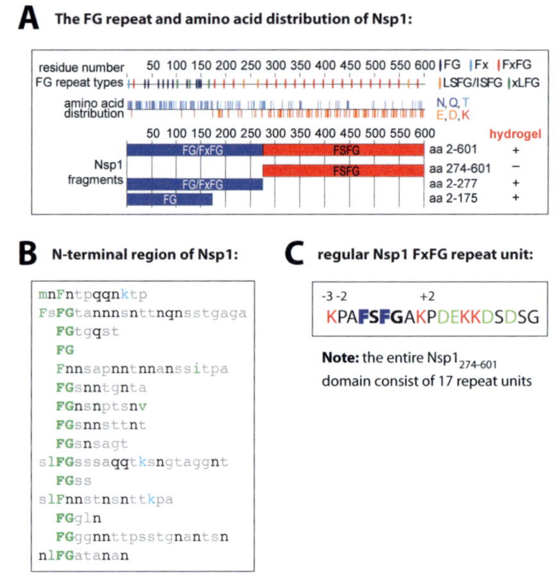

Figure 3.1. Overview of the Nsp1 FG domains. (A) Linear representation of non-uniform distribution of FG motifs and amino acids in Nsp1. Indicated below are the borders of the different Nsp1 FG domains based on the FG motif occurrence, amino acid composition and gelation propensity. Adapted from (Ader et al., 2010). (B) The amino acid sequence of the NQ-rich $Nsp1_{2-175}$ FG domain. FG motifs are highlighted in bold red, amido side chain resides in black and charged amino acids in blue. (C) Consensus sequence of the regular Nsp1 FxFG repeat unit. Hydrophobic residues are highlighted in bold blue, positively charged amino acids in red and negatively charged amino acids in green. The complete non-cohesive $Nsp1_{274-601}$ domain is comprised of seventeen FxFG repeat units, in which rarely single amino acids changes are found compared to the consensus sequence. Hence, for clarity the complete sequence is omitted.

2.2 The role of non-cohesive FG repeats for hydrogel-based NPC-like permeability barriers.

Please note that parts of the results described in this section have been published in (Ader et al., 2010), *as denoted in the figure legends.*

Despite the finding that cohesiveness seems to be a conserved feature of the Nup98-derived FG repeats, clearly not all known FG repeats are cohesive (Patel et al., 2007; Ader et al., 2010; Yamada et al., 2010). Hence, what is the functional role of such supposedly non-cohesive FG repeats?

A model FG repeat well suited to study this question is derived from Nsp1, an FG Nup localized to the central NPC channel of the baker's yeast *Saccharomyces cerevisiae*. As depicted in Figure 3.1, it contains (i) a very cohesive NQ-rich N-terminal FG repeat region ($Nsp1_{2-175}$) (Ader et al., 2010) and (ii) a highly regular, non-cohesive FSFG repeat region ($Nsp1_{274-601}$) (Ader et al., 2010; Yamada et al., 2010) with a high density of charged residues, which add up to a total net charge of +6.

2.2.1 The regular FSFG repeat region of Nsp1 passivates the hydrogel and facilitates the entry of active species.

In an attempt to elucidate the possible effect of the non-cohesive FG repeats on hydrogel-based permeability barriers, the active and passive selectivity properties of the N-terminal and full-length Nsp1 hydrogels were studied and compared. A yeast Impβ•IBB-MBP-GFP complex and MBP-mCherry were used as the active and passive permeation probes, respectively.

Two findings suggested that the charged $Nsp1_{274-601}$ FSFG repeats are crucial for the barrier performance of the full-length Nsp1 hydrogel. First, Impβ•IBB-MBP-GFP enriched in the outer areas of the $Nsp1_{2-175}$ hydrogel, but did not diffuse within it (Figure 3.2.A, upper panel). However, the same NTR•cargo complex was able to penetrate deep into the $Nsp1_{2-601}$ hydrogel (Figure 3.2.B, upper panel). Second, the $Nsp1_{2-175}$ hydrogel showed an unexpected affinity for MBP-mCherry, which enriched ~3-fold (Figure 3.2.A, lower panel). This interaction was fully suppressed in the $Nsp1_{2-601}$ hydrogel, i.e. when the charged FSFG repeats were present (Figure 3.2.B, lower panel).

Taken together, these findings suggest that Nsp1 acts like a two-component system, in which different requirements are met by distinct parts of the protein: the cohesiveness required to form a hydrogel-based permeability barrier is contributed by the NQ-rich N-terminus, which however needs to be 'passivated' by the highly charged FSFG repeats found in the C-terminal part of the FG repeat region.

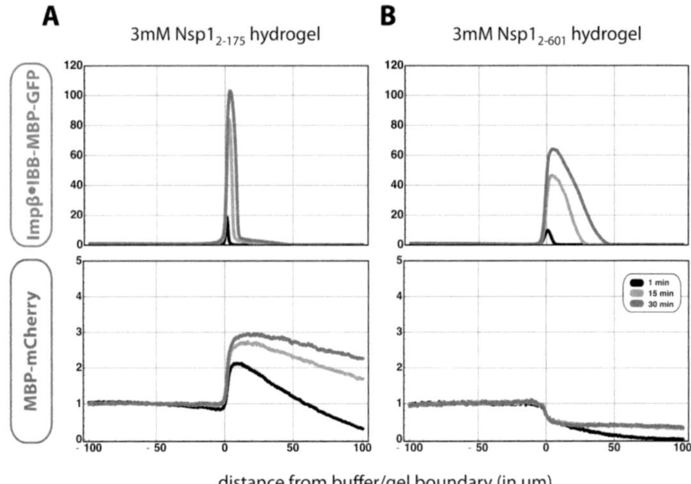

Figure 3.2. The charged Nsp1 FxFG repeat domain passivates the hydrogel against non-specific background binding of inert cargoes and facilitates the entry and intra-gel diffusion of NTR•cargo complexes. Hydrogels were formed with the indicated repeats at a concentration of 3mM and simultaneously challenged with 1µM of a yeast Impβ•IBB-MBP-GFP complex and 3µM MBP-mCherry. Influx of the fluorescent species was followed over time. Depicted here are the concentration profiles at t=1, 15 and 30 minutes. Note that the background binding of MBP-mCherry was completely reduced when the charged FSFG repeats were present. The intra-gel movement of the active species was about 3-times faster in the full-length Nsp1 hydrogel, whereas the partitioning coefficient was halved. These results have been published in (Ader et al., 2010).

2.2 The non-cohesive Nsp1 FSFG repeats also passivate the NQ-rich micronuclear *Tetrahymena* repeats to a certain degree.

Recall that also the hydrogels formed from the *Saccharomyces* and *Tetrahymena* Nup98-derived NQ-rich FG repeats show an unspecific background binding of MBP-mCherry (Figure 2.7). However, unlike Nsp1, they do not contain distinctive passivating regions in a *cis* configuration. Can the Nsp1$_{274-601}$ domain thus perhaps also optimize the performance of these barriers?

To test this idea, a tandem array of eighteen Nsp1-derived FSFG repeat units (Figure 3.1.C) was genetically fused to the extremely NQ-rich micronuclear *Tetrahymena* repeats (TtMicNup98A). The hybrid construct was purified under denaturing conditions (see Methods) and used to form hydrogels, which were subsequently challenged with Impβ•IBB-MBP-GFP and MBP-mCherry.

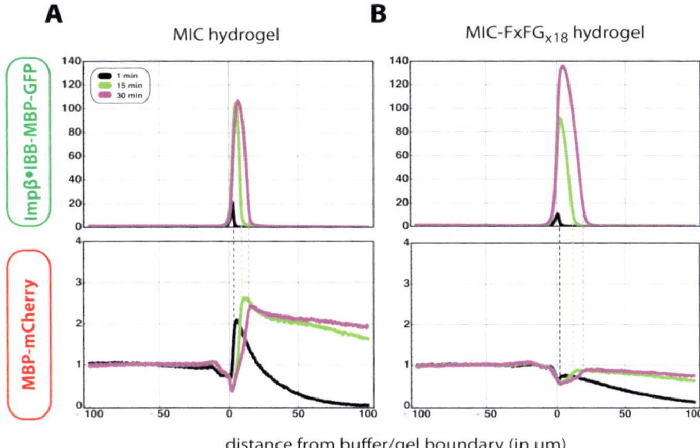

Figure 3.3. The effect of the charged Nsp1 FxFG repeats on the NQ-rich IF repeats from *Tetrahymena thermophila*. (A) Hydrogels formed from the wild-type TtMicNup98A repeats showed background binding of MBP-mCherry. Note that human Impβ•IBB-MBP-GFP was enriched ~50-times more in the gels than the passive species. (B) Hybrid hydrogels composed of a TtMicNup98A-FxFG$_{x18}$ fusion showed a drastically reduced MBP-mCherry binding, whereas the enrichment of the NTR•cargo complex was slightly stimulated. As indicated by the dashed line, MBP-mCherry was displaced in the parts of the gels were the active species accumulated. MBP-mCherry was used at 3μM and the Impβ•IBB-MBP-GFP complex at 1μM. The gels were casted at 200mg/ml.

The permeability of the TtMicNup98A and TtMicNup98A-FSFG$_{x18}$ hydrogels towards Impβ•IBB-MBP-GFP did not qualitatively differ (Figure 3.3, upper panels). Yet interestingly, the addition of the regular FSFG repeats caused a reduction of the partitioning coefficient of MBP-mCherry from ~2.5 to <0.9 (Figure 3.3, lower panels). However, in an ideal barrier, the expected partitioning coefficient for such molecules should be below 0.3. This suggests that in order to form highly selective hydrogels, FG repeats need to be meticulously fine-tuned.

2.3 Mutational analysis of the adhesive Nsp1$_{274-601}$ FSFG repeats.

The finding that the charged Nsp1$_{274-601}$ FSFG repeats are non-cohesive raised the question in how far they can nevertheless interact with the N-terminal Nsp1 FG repeats in the context of the hydrogel barrier. In the work of Ader et al. (2010) (where also the aforementioned results were published), this point was carefully addressed.

Remarkably, Nsp1$_{274-601}$ was indeed found to physically engage with Nsp1$_{2-175}$ in a specific manner, as it clearly enriched in an Nsp1$_{2-175}$ hydrogel, whereas an F-S mutant did not. Hence, the Nsp1$_{274-601}$ domain is certainly not 'cohesive' in the sense that it cannot form hydrogels on its own, but it is also neither truly 'non-cohesive'. Rather, it should be considered 'adhesive', i.e. able to specifically interact with other FG repeats.

Yet, is the interaction between the two FG repeat domains of Nsp1 however absolutely required for making a perfect barrier? If so, then a mutant version of the regular Nsp1$_{274-601}$ repeats that is no longer adhesive should not be able to passivate the NQ-rich Nsp1$_{2-175}$ domain anymore.

To create such a mutant, the amino acids constituting both the FSFG motifs of the Nsp1$_{274-601}$ domain (which were designated the signature sequence KPA**FSFG**AK here) and the intervening spacers were first systematically exchanged to identify allowed modifications that make the repeats more hydrophilic, but do not impair their ability to bind NTRs. The rules applied for mutagenesis were the following: (i) within groups of biochemically similar amino acids, the side chain with the fewest methyl group equivalents were favored (e.g. serine over threonine or arginine over lysine), (ii) amide residues were changed to aspartic acid and (iii) small aliphatic amino acids (e.g. alanine), as well as (iv) the cyclic imino acid side chain of proline were mutated to serine. Generally, neither the FG motif density, nor the net charge of the repeats (except for one mutant) were altered (see Figure 3.4.A for a complete overview of all mutants). Please note that short, Nsp1$_{176-601}$-derived model peptides comprising seven FG repeats each were used for the mutation analysis.

All mutants were obtained by *de novo* gene synthesis of the respective coding regions (see Materials). The nucleotide sequences were optimized for stability and expression in *E. coli*. All mutant repeats were affinity-purified under native conditions via N-terminal His- tags (see Methods). Equal amounts of the purified repeats were then immobilized on Ni^{2+}-beads again (see Figure 3.4.C and Methods) and used as baits to pull an Impβ-GFP•IBB-MBP-mCherry complex out of solution.

Interestingly, only three mutations affected NTR binding (see Figure 3.4.B and 3.4.D). First, the most severe effect was observed upon reversing the net charge of the FG repeats from +3 to – 3, which caused a marked reduction of NTR binding. Notably, only the charged residues in the spacer regions were exchanged, whilst leaving the positive charges flanking the FG motifs (and hence the local charge separation characteristic of Nsp1$_{176-601}$) untouched (see Mut2 in Figure 3.4.A). Second, it is noteworthy that NTR binding was enhanced when exchanging lysine, clearly the dominant charged amino acid in FG repeats, to the rarely occurring arginine side chain (see also Section 2.1.2). Third, NTR binding was reduced when the proline residues within the FG motifs were exchanged to serines (note however that this effect was not statistically significant). All other mutant FG repeats behaved similar to the WT. Importantly, this remains true when combining all 'silent' and the lysine to arginine mutations (see Mut8 in Figure 3.4.A).

Hence, the mutations comprising Mut8 were applied to the full length Nsp1$_{274-601}$ domain for further experiments.

A

	WT	Mut1	Mut2	Mut3	Mut4	Mut5	Mut6	Mut7	Mut8
Sequence	KPAFSFGAKPDENKASATS KPAFSFGAKPEEKKDDNSS KPAFSFGAKSNEDKQDGTA KPAFSFGAKPAEKNNNETS KPAFSFGAKSDEKKDGDAS KPAFSFGAKSDEKKDSDSS KPAFSFGTKSNEKKDSGSS	KPAFSFGAKPDENKSSSSS KPAFSFGAKPEEKKDDNSS KPAFSFGAKSNEDKQDGSS KPAFSFGAKPSEKNNNESS KPAFSFGAKSDEKKDGDSS KPAFSFGAKSDEKKDSDSS KPAFSFGTKSNEKKDSGSS	KPAFSFGAKPDEDKASATS KPAFSFGAKPEEKKDDDSS KPAFSFGAKSDEDKDDGTA KPAFSFGAKPAEKDDDETS KPAFSFGAKSDEKKDGDAS KPAFSFGAKSDEKKDSDSS KPAFSFGTKSDEKKDSGSS	KPAFSFGAKPDERKASATS KPAFSFGAKPEEKKDDDSS KPAFSFGAKSDEDKRDGTA KPAFSFGAKPAEKRDDETS KPAFSFGAKSDEKKDGDAS KPAFSFGAKSDEKKDSDSS KPAFSFGTKSREKKDSGSS	KPSFSFGSKPDENKASATS KPSFSFGSKPEEKKDDNSS KPSFSFGSKSNEDKQDGTA KPSFSFGSKPAEKNNNETS KPSFSFGSKSDEKKDGDAS KPSFSFGSKSDEKKDSDSS KPSFSFGSKSNEKKDSGSS	RPAFSFGARPDENKASATS RPAFSFGARPEEKKDDNSS RPAFSFGARSNEDKQDGTA RPAFSFGARPAEKNNNETS RPAFSFGARSDEKKDGDAS RPAFSFGARSDEKKDSDSS RPAFSFGTRSNEKKDSGSS	RPSFSFGSRPDENKASATS RPSFSFGSRPEEKKDDNSS RPSFSFGSRSNEDKQDGTA RPSFSFGSRPAEKNNNETS RPSFSFGSRSDEKKDGDAS RPSFSFGSRSDEKKDSDSS RPSFSFGSRSNEKKDSGSS	KSSFSFGSKSDENKASATS KSSFSFGSKSEEKKDDNSS KSSFSFGSKSNEDKQDGTA KSSFSFGSKSAEKNNNETS KSSFSFGSKSDEKKDGDAS KSSFSFGSKSDEKKDSDSS KSSFSFGSKSNEKKDSGSS	RPSFSFGSRPDERRSSSSS RPSFSFGSRPEERRDDDSS RPSFSFGSRSDEDRRDGSS RPSFSFGSRPSERRDDESS RPSFSFGSRSDERRDGDSS RPSFSFGSRSDERRDSDSS RPSFSFGSRSRERRDSGSS
Mutations		*Within Spacer Region:* A/T-S	*Within Spacer Region:* N/Q-D	*Within Spacer Region:* N/Q-D/R (i.e. charge neutral)	*Within FSFG Cluster:* A-S	*Within FSFG Cluster:* K-R	*Within FSFG Cluster:* A-S K-R	*Within FSFG Cluster:* A-S P-S	*Within Spacer Region:* A/T-S K-R N/Q-D/R *Within FSFG Cluster:* A-S K-R
FG clusters/ 100AS	5.26	5.26	5.26	5.26	5.26	5.26	5.26	5.26	5.26
charges/ 100AS	3	3	-3	3	3	3	3	3	3
hydro-phobicity	-1.268	-1.368	-1.268	-1.298	-1.523	-1.331	-1.586	-1.462	-1.765

B

C

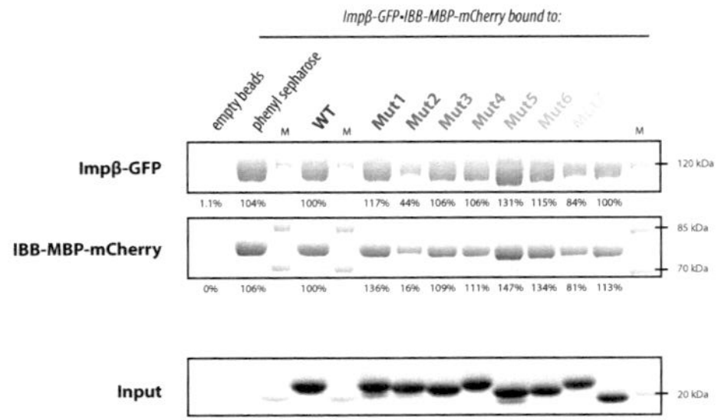

D

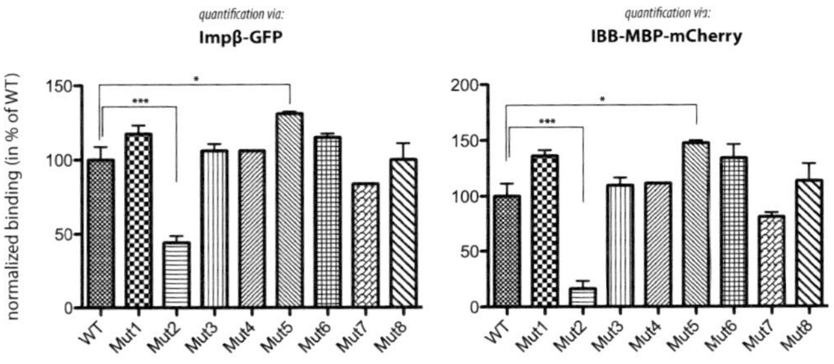

Figure 3.4. The hydrophobicity of the Nsp1$_{274-601}$ domain can be reduced without loosing its capability to efficiently bind NTRs (previous page). (A) In order to reduce the adhesiveness of the FxFG repeat domain, a series of mutations was introduced that effected neither its FG motif, nor charge density, but reduced its overall hydrophobicity. Note that only an Nsp1$_{274-601}$-derived, seven FxFG repeat unit domain was systematically mutated and tested for NTR binding, as gene synthesis of the corresponding constructs was thereby greatly facilitated. The mutants were expressed in *E.coli* and purified under denaturing conditions, as described in the Methods. (B,C) To estimate their NTR binding capabilities, ~20-30µg of the mutants were immobilized o/n to a Ni^{2+} 500Å silica matrix and incubated with 5µM of a human Impβ-GFP•IBB-MBP-mCherry complex for 1 hour at RT. Afterwards, the GFP and mCherry fluorescence in the buffer was quantified and the amount of the NTR•cargo complex bound to the respective mutants calculated. Given is the relative enrichment of the NTR•cargo complex on the mutants normalized to the wild-type repeats. (D) Statistical analysis of the binding experiment, which comprised three independent replicas of each binding reaction. In comparison to the WT, the Mut2 repeats showed a significantly reduced affinity for the NTR•cargo complex (i.e. $P<0.05$ in a Newmann-Keuls test), whereas Mut5 bound the complex significantly better. Note that the quantification and statistical analysis was consistent independent of the fluorescent tag used for quantification.

As expected, the Nsp1$_{274-601}$^{Mut8} repeats (like the WT Nsp1$_{274-601}$ domain) does not form hydrogels. To test whether it still interacts with the gel-forming Nsp1 portion (i.e. Nsp1$_{2-277}$), the different FG domains of Nsp1 were coupled to Ni^{2+}-Silica beads via N-terminal His-tags and used to pull a GFP-labelled version of Nsp1$_{2-277}$ out of solution.

In agreement with published results (Ader et al., 2010), the cohesive Nsp1$_{2-277}$ domains interacted strongly with each other: ~90% of the mobile variant were bound by the immobilized bait (Figure 3.5). The Nsp1$_{274-601}$ domain trapped the Nsp1$_{2-277}$-GFP baits less efficiently, binding only ~70% of the input. The immobilized Nsp1$_{274-601}$^{Mut8} repeats however only bound approximately half of the total Nsp1$_{2-277}$-GFP amount put in.

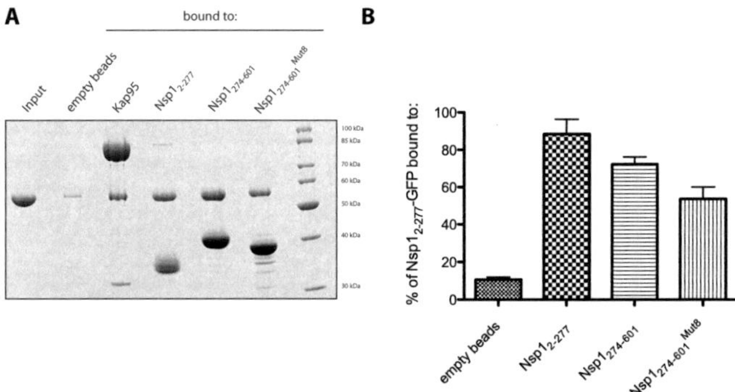

Figure 3.5. The adhesiveness of the Nps1$_{274-601}$ domain can be reduced by making it more hydrophilic. (A) The mutations comprising Mut8 were applied to the full-length Nsp1$_{274-601}$ domain and 30µg of this, as well as the Nsp1$_{2-277}$ and wild-type Nsp1$_{274-601}$ domains were immobilized o/n to a Ni^{2+} 500Å silica matrix. After incubation with 5µM of a GFP-labelled Nsp1$_{2-277}$ domain for 4h at RT, the remaining GFP fluorescence in the supernatant was measured and the relative amount of Nsp1$_{2-277}$-GFP bound to the indicated species calculated. For each binding reaction, three replicas were performed. (B) Statistical

analysis revealed that the wild-type $Nsp1_{274-601}$ domain, despite being unable to form hydrogels on its own and thus termed 'non-cohesive', did not interact significantly less with the $Nsp1_{2-277}$ prey compared to the highly cohesive $Nsp1_{2-277}$ domain. In contrast, the $Nsp1_{274-601}^{Mut8}$ domain indeed was significantly less efficient in pulling the prey out of solution when compared to the cohesive $Nsp1_{2-277}$ domain. Note however that an interaction is nevertheless clearly observed.

2.4 The $Nsp1_{274-601}$ FSFG repeats seem to be highly optimized sequences.

Interestingly, exchanging the WT $Nsp1_{274-601}$ domain with the $Nsp1_{274-601}^{Mut8}$ repeats nevertheless severely influenced the permeability properties of the Nsp1 hydrogel (Figure 3.6, B and C). Indeed, the $Nsp1_{2-175}$-$Nsp1_{274-601}^{Mut8}$ gel behaved similar to an $Nsp1_{2-175}$ gel: first, it also enriched MBP-mCherry and second, it did not allow an efficient intra-gel diffusion of Impβ-IBB-MBP-GFP (Figure 3.6, A and C).

In comparison, hydrogels formed from Nsp1 repeats in which the FSFG motifs of the $Nsp1_{274-601}$ domain were converted into SSSG themes, thereby eliminating NTR binding sites (Bayliss et al., 2000; Frey et al., 2006) and abolishing the adhesiveness of this region (Ader et al., 2010), did not significantly enrich Impβ-IBB-MBP-GFP, but remarkably showed a reduced MBP-mCherry binding (Figure 3.6 D). The same was true for hydrogels formed from a fusion of the $Nsp1_{2-175}$ domain and a $(GTGSSGNSGSGTGSSGNSGS)_{x19}$ repeat likewise lacking adhesiveness and the ability to bind NTRs (Figure 3.7 E).

Taken together, these findings suggest that the $Nsp1_{274-601}$ FSFG repeats are highly optimized sequences sensitive to subtle changes, even when these do not directly affect *bona fide* NTR binding sites (but e.g. their hydropathy and/or adhesiveness).

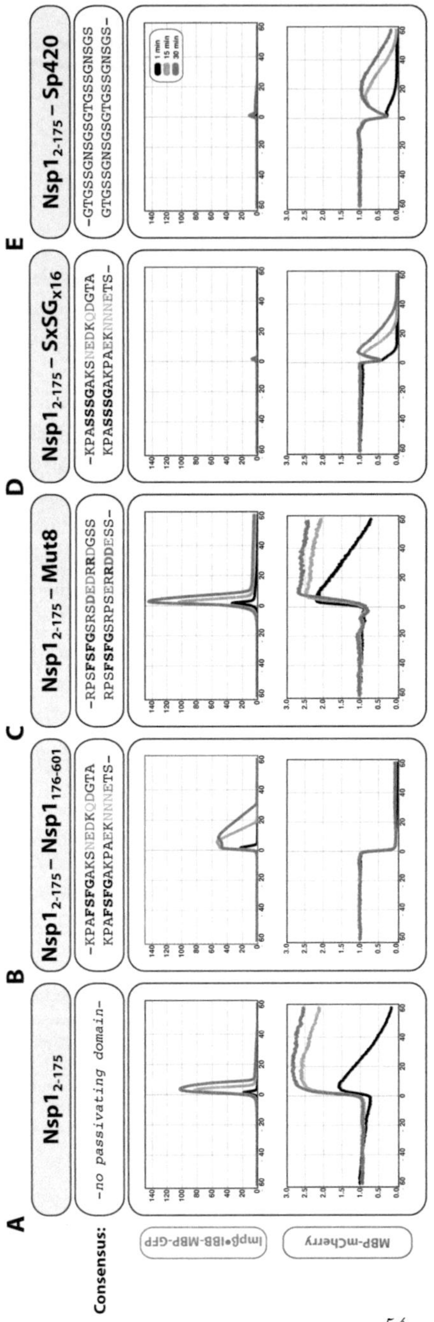

distance from buffer/gel boundary (in µm)

Figure 3.6. Despite its remaining interaction potential with cohesive FG repeats, the Nsp1$_{274-601}$^{Mut8} domain cannot passivate the Nsp1$_{2-175}$ hydrogel like the wild-type Nsp1$_{274-601}$ repeats. Hydrogels were formed from the indicated constructs at 3mM each and simultaneously incubated with yeast Impβ•IBB-MBP-GFP (1µM) and MBP-mCherry (3µM) permeation probes. The concentration profiles of both species after t=1, 15 and 30 minutes of influx in the indicated gels are given. Depicted as Consensus are two of the 16-19 repeat units characteristic of each domain fused to Nsp1$_{2-175}$ for passivation, respectively. (C) Note that the Nsp1$_{274-601}$^{Mut8} domain does not reduce the background binding of MBP-mCherry. (D) The SxSG$_{x16}$ repeats are an Nsp1$_{274-601}$-derived domain in which all phenylalanine residues have been mutated to serine. This mutation dramatically reduced the adhesiveness of the domain (Ader et al., 2010) and abolished NTR binding. (E) Sp420 denotes a 420 amino acid-long spacer comprised mainly of glycine, serine and threonine residues, which lacks any hydrophobic clusters and hence neither is adhesive, nor binds NTRs. Both of these domains passivate the Nsp1$_{2-175}$ hydrogel to a certain degree, but almost completely abolish NTR binding. See main text for discussion.

2.3 Conversion of the prototypical NQ-rich yeast prion Sup35 into a functional FG repeat.

Proteins in general have a strong tendency to form amorphous aggregates upon denaturation. This is largely due to the association of the hydrophobic segments normally buried in the folded protein cores (see e.g. Jaenicke and Seckler, 1997). Some proteins however readily assemble into amyloids, the latter being ordered, fibrillar aggregates stabilized by extensive interchain β-sheets (Sunde and Blake, 1997). Prominent examples are prions, which are capable of forming amyloids in a self-perpetuating, infectious and (non-Mendelian) inheritable manner (Chien et al., 2004).

Arguably, most prions have been found in the yeast *Saccharomyces cerevisiae* (see e.g. Halfmann et al., 2010 for review). Amyloid formation of yeast prions is mediated by modular and transferable prion-determining domains (PrDs) (Li and Lindquist, 2000; Alberti et al., 2009). Yeast PrDs are highly NQ-rich (DePace et al., 1998; Toombs et al., 2010), a characteristic feature that has been used to identify more than 200 prion-like yeast proteins (Michelitsch and Weissman, 2000; Alberti et al., 2009). Notably, amongst the many hits were several FG Nups.

Indeed, it was recently shown that the GLFG repeat region from Nup100 behaves as a prion under certain conditions (Halfmann et al., 2011). Moreover, ssNMR studies of the $Nsp1_{2-601}$ hydrogel revealed amyloid-like interchain β-sheet formation between the NQ-rich portions of the FG repeats. These findings hence raise the questions in how for prions and FG repeats are related and whether *bona fide* amyloid-forming proteins such as yeast prions can be converted into functional, barrier-forming FG repeats.

2.3.1 The yeast translation termination factor Sup35.

The PrD of the yeast translation termination factor Sup35 (a homolog of the eukaryotic release factor eRF3) is a prototypical NQ-rich prion. It spans approximately the first sixth of the protein and is extremely rich glutamine and asparagine, as well as tyrosine (Figure 4.1.A). Under normal physiological conditions, the PrD is dispensable for Sup35 to function. Separating the PrD from the catalytically active EF1 domain[2] is the middle (M) domain, which contains many charged residues (Figure 4.1.A).

---

[2] The crystal structure of the Sup35 EF-domain from *Schizosaccharomyces pombe* has recently been solved and shown to bear structural resemblance to eukaryotic translation elongation factor 1α (EF1α) and the prokaryotic elongation factor EF-Tu (Kong et al., 2004). It possesses GTPase activity and binds to the termination factor Sup45 (the yeast homolog of eRF1) (Stansfield et al., 1995; Zhouravleva et al., 1995), the latter of which mimics tRNA and fits to the ribosomal A-site (Cheng et al., 2009). There, Sup45/eRF1 recognizes the stop codon and catalyzes the hydrolysis of the ester bond between the tRNA and the bound polypeptide chain, thus releasing newly synthesized proteins from the ribosome. Importantly, the translation termination activity of Sup45/eRF1 is stimulated by Sup35/eRF3 in a GTP-dependent manner (Zhouravleva et al., 1995; Salas-Marco and Bedwell, 2004). Thus, when Sup35 is converted to its amyloid form, translation termination deceases, as the EF1 domain is no longer accessible to the ribosome (see also Section 1.4).

Although the M domain has been implicated with amyloid formation as well (e.g. in Shewmaker et al., 2009), only the first 140 amino acids, which feature the prototypical amino acid composition of yeast amyloid-forming proteins (Michelitsch and Weissman, 2000; Alberti et al., 2009) and include the PrD, were chosen as the model prion for this study.

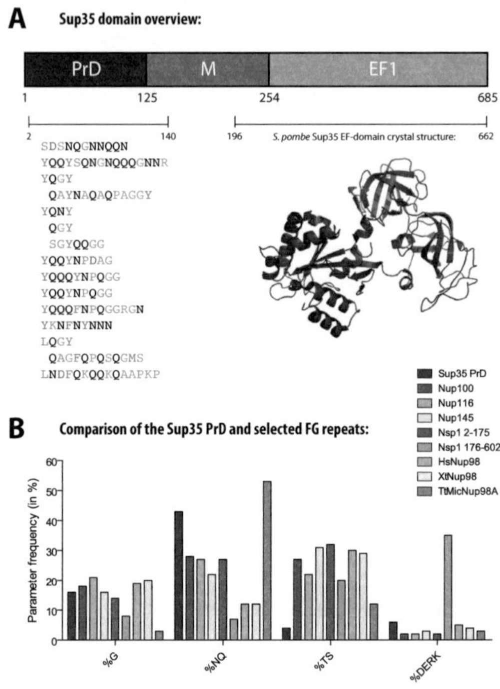

Figure 4.1. The yeast translation termination factor Sup35. (A) Sup35 comprises three domains: the NQ-rich prion-determining domain (PrD), the middle domain (M) and the catalytically active EF domain. (B) The amino acid sequence of the Sup35 PrD shows that tyrosine is the predominant hydrophobic residue. Comparison of the Sup35 PrD to selected FG repeats (see legend). Note that the Sup35 PrD is relatively enriched in NQ, but depleted of TS.

2.3.2 Sequence comparison between NQ-rich FG repeats and the Sup35 PrD.

Notably, NQ-rich FG repeats and the PrDs of yeast prions do not only share the same key structural elements in their aggregated forms (Ader et al., 2010), but also show a similar amino acid bias: both kinds of proteins are enriched for glycine and polar residues such as asparagine and glutamine, whereas they are largely depleted of charged and most hydrophobic amino acids. An exception to the latter observation is in both cases usually a single dominant aromatic amino acid, which in turn is oftentimes enriched (see also Figure 2.2.A). Amongst the many (yeast) FG repeats, the Sup35 PrD is particularly comparable to the N-terminal part of Nsp1 (Nsp1$_{2-175}$), as they are almost identical in length (140 and 175 amino acids, respectively), contain the same overall net charge (+3) and, considering the variance observed for FG repeats in this parameter, are also similarly hydrophilic (0.4 and 0.8, respectively).

Despite these general similarities, clear differences between the Sup35 PrD and FG repeats are evident (see Figure 4.1.B). Firstly, the overall NQ-content of the Sup35 PrD (nearly 45%) is significantly higher compared to selected FG repeats (below 30% even for the NQ-rich Nsp1$_{2-175}$ or the yeast Nup98 homologs Nup100, Nup116 and Nup145). An exception are the micronuclear *Tetrahymena* Nup98 repeats, which contain more than 50% NQ. In this regard however, it should secondly be noted that glutamine (29%) is more prevalent than asparagine (15%) in the Sup35 PrD, whereas the opposite is true for FG repeats, especially the micronuclear *Tetrahymena* examples (see Section 2.1.2). Thirdly, the dominant hydrophobic residue is clearly phenylalanine in the case of FG repeats (see Figure 2.2.A), yet tyrosine for the Sup35 PrD. Finally, the Sup35 PrD is strikingly devoid of serine and threonine residues, which are highly abundant in FG repeats (Figure 4.1.B; see also Figure 2.2.A).

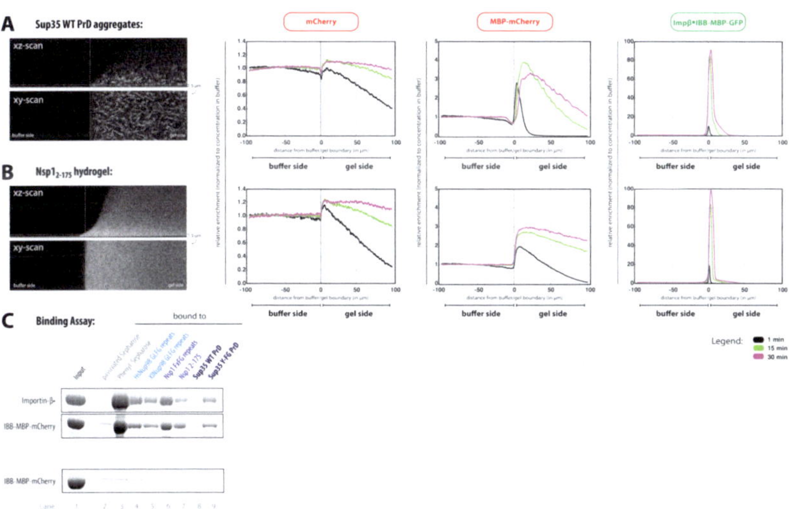

Figure 4.2. The Sup35 PrD forms fibrillar aggregates under conditions that NQ-rich FG repeats from hydrogels. The Sup35 PrD and Nsp1$_{2-175}$ domain were expressed and purified under identical conditions. The lyophilized proteins were re-suspended at 200mg/ml in 0.2% TFA (see Methods). Whereas the Nsp1$_{2-175}$ domain formed homogenous hydrogels, the Sup35 PrD formed characteristic fibrillar aggregates. (A,B) The Sup35 PrD aggregates and Nsp1$_{2-175}$ hydrogels observed with a laser scanning confocal microscope and the partitioning of mCherry (used at 3μM), MBP-mCherry (at 3μM) and yeast Impβ•IBB-MBP-GFP (at 1μM) into the two different structures. Shown are the concentration profiles for t = 1, 15 and 30 minutes of influx. Note that both the Sup35 PrD aggregates and Nsp1$_{2-175}$ hydrogels show similar permeability properties; see

main text for discussion. (C) Binding of IBB-MBP-mCherry, alone or in complex with Impβ, to the indicated FG repeats or Sup35 PrD constructs. The FG repeats have been covalently coupled at identical concentrations to a Sepharose matrix (see Methods). Note that the NTR•cargo complex binds to the Nsp1$_{2-175}$ domain, but not to the Sup35 PrD. However, after introduction of *bona fide* FG motifs, the Sup35 PrD is capable of interacting with NTRs (see also Section 2.3.5).

2.3.3 The Sup35 PrD forms fibrillar aggregates that bind inert molecules and NTRs.

Notably, both the Nsp1$_{2-175}$ FG domain and the Sup35 PrD were cohesive under the standard conditions used for hydrogel formation in this study (i.e. resuspension of the lyophilized proteins and gelation in 0.2% TFA). However, the Sup35 PrD reproducibly formed water-poor, fibrillar aggregates (Figure 4.2.A), rather than water-swollen, elastic hydrogels as the Nsp1$_{2-175}$ FG domain (Figure 4.2.B).

These findings reflect the conceptual difference between hydrogels and amyloids, namely that hydrogels (by definition) contain a high degree of solvent, whereas water molecules tend to be excluded in amyloid aggregates as a result of the dense subunit packing within the fibrils (Chiti and Dobson, 2006). Clearly, this is also highly relevant for the NPC, as it is hard to conceive how such fibrils could form an adequate selectivity barrier.

Hence, it was surprising to find that the Sup35 PrD aggregates nevertheless showed an affinity for a yeast Impβ•IBB-MBP-GFP complex (Figure 4.2.A), which actually was qualitatively comparable to the interaction between the Nsp1$_{2-175}$ FG hydrogel and the same NTR•cargo complex (Figure 4.2.B). Moreover, both the Sup35 PrD aggregates and the Nsp1$_{2-175}$ FG hydrogel bound passive permeation probes such as mCherry or MBP-mCherry (Figure 4.2.A and B).

Differences in the affinity of the Sup35 PrD and the Nsp1$_{2-175}$ domain for an Impβ•IBB-MBP-mCherry complex were first observed in a classical bead-binding assay, which is less sensitive than the hydrogel system. In contrast to the Nsp1$_{2-175}$ fragment, the Sup35 PrD did not bind the NTR•cargo complex when both were coupled to Sepharose beads (in a much lesser concentration compared to the hydrogels/aggregates) (Figure 4.3.C). This finding indicates that, as expected, the interaction between the prion and Impβ is indeed weaker than the interaction between the *bona fide* FG repeats and the NTR. Neither the immobilized PrD, nor the FG repeat bound MBP-mCherry (Figure 4.3.C)

Taken together, two major challenges for converting a *bona fide* prion into a fully functional FG repeat were identified: (1) preventing the aggregation of the Sup35 PrD into amyloid-like fibrils and (2) introducing stronger NTR binding sites.

2.3.4 The exchange of all tyrosines into phenylalanines allows mutant Sup35 PrDs to form homogenous hydrogels.

What could the simplest and most straightforward change with the highest impact on converting the Sup35 PrD into a functional FG repeat be? A striking difference between both kinds of proteins is that tyrosine, instead of phenylalanine, is the favored hydrophobic amino acid in the Sup35 PrD. Phenylalanine residues however play a crucial role in FG repeats, as they are essential for both NTR binding (being integral parts of FG motifs; Bayliss et al., 2000; 2002b) and hydrogel formation (F to S mutants of FG repeats fail to form hydrogels; Frey et al., 2006). Hence, changing all tyrosines of the Sup35 PrD into phenylalanines (yielding the mutant Sup35 Y-F PrD; see Figure 4.3.B) seemed plausible. In addition, another mutant was designed by rearranging the Sup35 Y-F PrD in such a way that true FG clusters were created (see Figure 4.3.B). This Sup35 Y-FG PrD indeed had an improved affinity for Impβ•IBB-MBP-mCherry in comparison to the Sup35 WT PrD, as it was able to bind the NTR•cargo complex even when immobilized to Sepharose beads (Figure 4.2.C).

Unexpectedly, already the Y-F mutation had a number of drastic effects. First, the Sup35 Y-F PrD formed homogenous, transparent hydrogels instead of the fibrillar aggregates observed for the wild-type PrD (see Figure 4.3.A). Second, the Sup35 Y-F PrD hydrogel allowed the efficient entry and intra-gel diffusion of an Impβ•IBB-MBP-GFP probe, as indicated by the high enrichment (~120-fold) and penetration depth (~40µm within 30 minutes) of the NTR•cargo complex (Figure 4.3.D). Third, the Sup35 Y-F PrD gel showed signs of a sieve-like passive barrier, as a monomeric mCherry (~30kDa) and a tetrameric tCherry (~120kDa) species were hindered to enter the gel in a size-dependent manner (Figure 4.3.C). This is perhaps best appreciable when taking a look at the partitioning coefficient of the two species 100µm deep into the gel at t=1min, which is already ~0.3 for mCherry and only <0.1 for tCherry. After t=30min, partitioning coefficients of ~0.9 and ~0.75 were observed for mCherry and tCherry, respectively (Figure 4.3.C).

The Sup35 Y-FG PrD also formed hydrogels (Figure 4.3.A), which behaved similar towards Impβ•IBB-MBP-GFP (Figure 4.3.D) and MBP-mCherry (Figure 4.3.C) as the Sup35 Y-F hydrogel. The only differences were that (i) the enrichment of the NTR•cargo complex was lower (80 vs. 120, respectively) and (ii) the partitioning coefficients observed for mCherry and tCherry were slightly higher (~1.0 vs. ~0.9/~0.75, respectively).

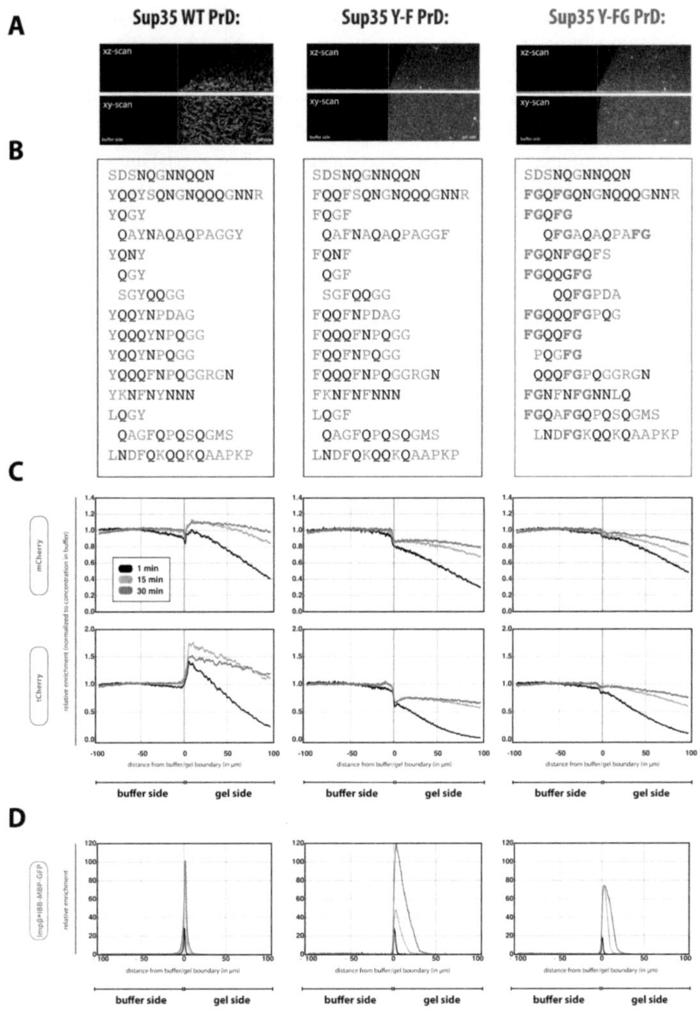

Figure 4.3. Minimal changes to the Sup35 PrD are sufficient to convert it into an FG repeat-like domain that forms selective hydrogels. All proteins were re-suspended at 200mg/ml in 0.2% TFA. (A) A mere exchange of tyrosine to phenylalanine allowed the formation of homogenous hydrogels. (B) Rearrangement of the Sup35 Y-F into a Sup35 Y-FG domain. (C) In contrast to the Sup35 WT PrD, the Sup35 Y-F (and to a lesser degree the Sup35 Y-FG PrD) forms hydrogels that restrict the influx of inert molecules such as mCherry or tCherry. (D) The introduced mutation also stimulated the influx and intra-gel diffusion of yeast Impβ•IBB-MBP-GFP. Strikingly, this effect was more pronounced for the Sup35 Y-F PrD hydrogel, despite its lack of *bona fide* NTR binding sites. Shown here are the concentration profiles of the indicated species after t = 1, 15 and 30 minutes of influx. As usual, the active species was used at 1μM and the passive species at 3μM.

2.3.5 The Sup35 Y-FG PrD binds NTRs from HeLa extracts.

Further support for the notion that the Sup35 Y-FG PrD can specifically bind NTRs came from the remarkable finding that it could, when immobilized to Sepharose beads, pull-out a plethora of importins, import adaptors, exportins and even nucleoporins from HeLa extracts, as identified by subsequent mass spectrometry analysis (Figure 4.4). Amongst them were Impβ and Crm1, the major import and export factor, respectively. In fact, from the reported NTRs in the literature (see Table 1.1), only the importins Imp4, Imp9, Imp11 and Imp13, as well as the exportins Exp4 and Exp-t were not found.

The most prominent nucleoporin hit in the dataset (and actually one of the most prominent hits overall) was the inner NPC ring component Nup93. Interestingly, the crystal structure of its yeast homolog Nic96 was recently solved and revealed that, except for the N-terminal coiled-coil region involved in anchoring the Nsp1/Nup62-complex (see Section 1.2), the majority of the protein bears surprisingly great structural similarity to Impβ (Schrader et al., 2008). In the same paper, a Nic96 version lacking its coiled-coil region was shown to bind FG repeats *in vitro*. Thus, a plausible explanation for the presence of Nup93 and other Nup93-Nup205 complex components in the mass spectrometry dataset (as well as Nups belonging to associated complexes) is that the Sup35 Y-FG PrD physically interacted with the Impβ-like part of Nup93 and thereby pulled the respective factors out of the cell lysate.

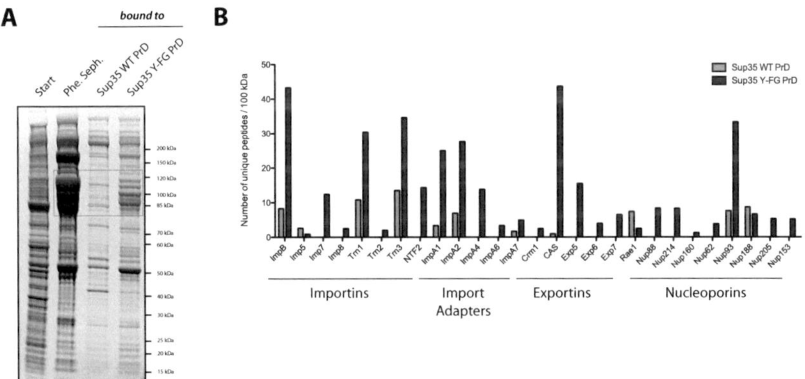

Figure 4.4. The Sup35 Y-FG PrD binds to a plethora of NTRs from a HeLa cell extract. (A) The Sup35 WT and Y-FG PrD were covalently coupled to a Sepharose matrix and incubated with HeLa extracts (kind gift of P. Odenwälder) for 2 hours at 4°C (see Methods). As a positive control for NTR binding, a phenyl sepharose matrix was used (Ribbeck and Görlich, 2002). Highlighted in the dashed box is the molecular weight range of NTRs (except for NTF2). The pull-downs were subsequently analyzed with mass spectrometry. (B) For the qualitative comparison of the mass spectrometry results, the number of unique peptides found for a given hit was normalized to its molecular weight. Compared here are all identified NTRs (grouped into importins, import adaptors and exportins) and FG repeats that were bound by the two Sup35 PrD constructs. Note that in contrast to the Sup35 WT PrD, the Sup35 Y-FG PrD binds almost all NTRs (see also Table1.1).

Notably however, also the unmodified Sup35 WT PrD, when analogously immobilized, was able to pull-out a few factors, namely the importins Impβ (and with it Impα1, Impα2 and Impα7), Trn and Imp5, as well as the mRNA export factor Rae1 and the nucleoporins Nup93 and Nup188. Surprisingly, the MW-normalized number of unique peptides found for Imp5 and Rae1 was even greater in the Sup35 WT PrD sample compared to the Sup35 Y-FG PrD sample.

2.3.6 The different Sup35 PrDs can be passivated by the Nsp1$_{176-601}$ domain to form selective hydrogels with NPC-like permeability properties.

Despite the success of transforming the Sup35 PrD into hydrogel-forming and NTR-binding domains, the obstacle remains that the mutants did not constitute tight barriers with NPC-like sieving effects. This however also seems to be a problem of certain NQ-rich yeast FG repeats, as e.g. shown in this study for the Nup145-derived GLFG repeats (see Figure 2.7) or the Nsp1$_{2-175}$ FG domain (see Figure 4.2.B). In latter case, the charged Nsp1$_{274-601}$ FSFG repeats apparently optimize the barrier properties of the full-length Nsp1$_{2-601}$ FG/FSFG domain (see Section 2.2). Could the Nsp1$_{274-601}$ FSFG repeats hence perhaps have a similar effect on the Sup35 PrDs?

Remarkably first, the Sup35 WT PrD did no longer form fibrillar aggregates when fused to a Nsp1$_{176-601}$ domain. However, the fusion protein neither formed homogenous hydrogels, possibly reflecting the strong self-association tendency of the Sup35 WT PrD (Figure 4.5A). In contrast, the Sup35 Y-FG PrD-Nsp1$_{176-601}$ fusion assembled into a homogenous hydrogel phase (Figure 4.5.B), which was almost identical in its appearance to the wild-type Nsp1 hydrogel (Figure 4.5.C).

Despite their morphological differences, all three hydrogels showed strikingly similar permeability properties: the partitioning coefficient for MBP-mCherry was around 0.3, whereas Impβ•IBB-MBP-GFP enriched on average nearly 100-fold and within 30 minutes penetrated roughly 50µm deep into the hydrogel (Figure 4.5). Thus, in comparison to the WT PrD or Nsp1$_{2-175}$ FG domain, the partitioning coefficient of MBP-mCherry was effectively lowered by a factor of ~10 and the penetration depth of Impβ•IBB-MBP-mCherry increased ~5-fold.

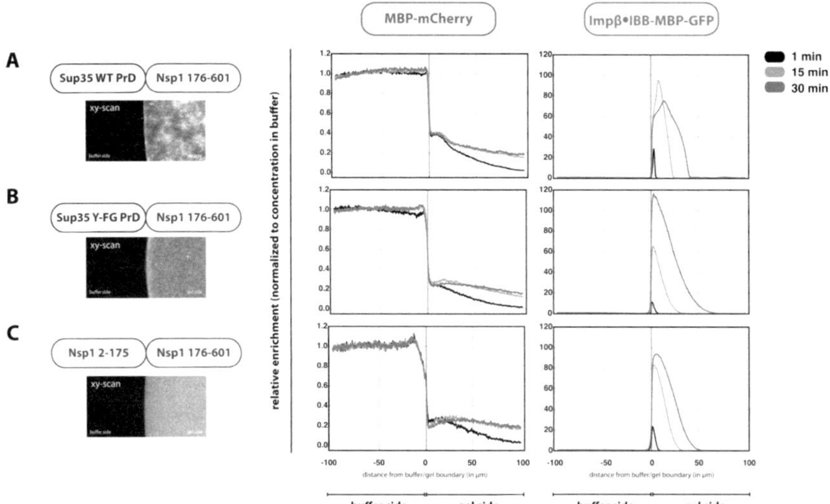

Figure 4.5. The regular Nsp1 FxFG repeats can optimize the performance of the Sup35 PrD hydrogels. All gels were formed at 200mg/ml and challenged with 1µM of the indicated active and 3µM of the passive species. The concentration profiles for the influx of the indicated probes after t = 1, 15 and 30 minutes are given. (A) Note that a fusion of the Sup35 WT PrD with the FxFG repeats does not form fibrillar aggregates, but inhomogenous hydrogels. The presence of the charged Nsp1 domain passivates the Sup35 WT PrD against the binding of MBP-mCherry and facilitates the influx and intra-gel diffusion of yeast Impβ•IBB-MBP-GFP. (B) The FxFG repeats have a similar effect on the Sup35 Y-FG PrD. (C) For comparison, a full-length Nsp1 hydrogel was challenged with the same permeation probes. Note that there are no qualitative changes between the performances of the three gels.

2.4 Reconstitution of the orthogonal NPC permeability barriers of the bi-nucleated ciliate *Tetrahymena thermophila*.

Hitherto, the major focus of the work presented here was directed at the question how FG repeats shape the NPC permeability barrier. In particular, the ability of FG repeats from different eukaryotic clades to form a selective phase, i.e. FG hydrogels (Part I of the results section), and the therefore required sequence and structural features (Parts II and III) were systematically evaluated. Amongst other findings, it was shown that (i) all tested FG repeats indeed formed hydrogels and interestingly, (ii) human Impβ was able to interact with all of these, despite the evolutionary distance and sequence divergence between the individual FG repeats constituting the hydrogel barriers (see Figure 2.7). The latter finding highlights the paramount challenge of nuclear dualism faced by bi-nucleated ciliates such as *Tetrahymena thermophila*: the requirement of adapted nuclear transport apparatuses, comprising at least (i) two different kinds of permeability barriers with orthogonal selectivity and (ii) nucleus-specific NTRs (see also Introduction).

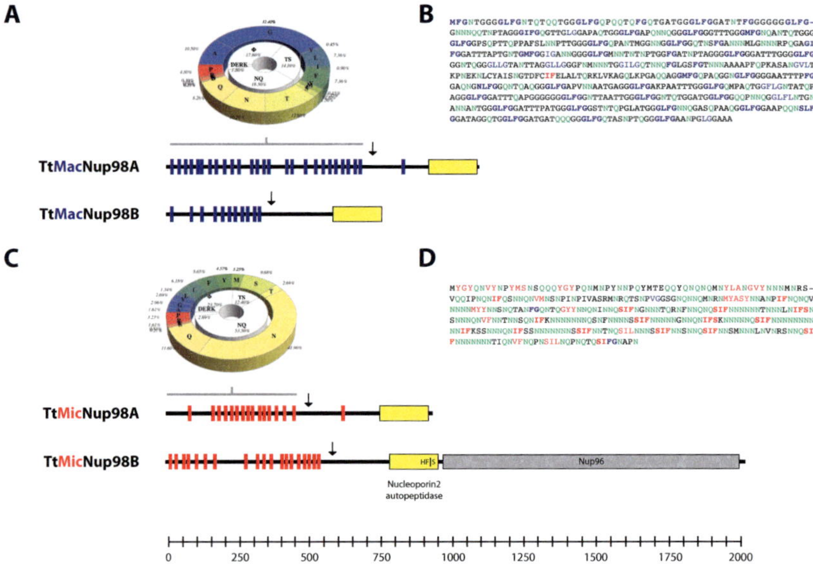

Figure 5.1. Comparison of the Nup98 homologs of *Tetrahymena thermophila*. The macro- and micronuclear NPC permeability barriers of the unicellular eukaryote *Tetrahymena* contains are guarded by different Nup98 homologs, of which two localize exclusively to the MAC or MIC, respectively. (A) Schematic overview of the macronuclear Nup98 homologs. Blue bars represent GLFG motifs and the yellow

box the nucleoporin2 autopeptidase domain (see also Section 2.1.1). The arrows indicate the parts designated as the FG repeats regions in the present study. Above the FG repeat region is a graphical representation of the amino acid composition of this domain. Shown in (B) is the amino acid sequence of the TtMacNup98A GLFG repeat region. (C) Overview of the micronuclear Nup98 homologs. Red bars indicate IF motifs, the yellow box represents the nucleoporin2 domain. Note that only TtMicNup98B is present as a fusion protein to Nup98. In agreement with this organization, only the nucleoporin2 domain of TtMicNup98B contains the catalytic HF|S site. (D) Amino acid sequence of the IF repeat region from TtMicNup98A.

The unique cell biology of ciliates thus raises the question how the concepts introduced and investigated up to this point are refined to operate two morphologically and functionally distinct nuclei in a common cytoplasm. Specifically, how must FG repeats differ to allow the formation of permeability barriers with orthogonal selectivity properties?

Arguably the most straightforward solution would be to have subsets of FG repeats with fundamentally different NTR binding sites (i.e. FG motifs), which can exclusively be recognized by specialized NTRs. Indeed, it was shown that the macro- (MAC) and micronucleus (MIC) posses strikingly dissimilar Nup98 orthologs (see Figure 5.1), which either contain prototypical GLFG or unique IF motifs, respectively (Malone et al., 2008; Iwamoto et al., 2009). Notably, within the Nup98-derived FG repeat database compiled in this study, which includes sequences from all eukaryotic clades, such IF motifs were only found in a subset of repeats from *Tetrahymena* and *Paramecium*[3] (i.e. the MIC ones). In the cluster analysis of the database, the ciliate IF repeats composed a very defined, closed subgroup, further highlighting their relative disparity in comparison to all other Nup98 homologs (see Section 2.1.3).

2.4.1 A mere exchange of FG motifs is not sufficient to alter the specificity of macronuclear FG repeats.

What would hence happen if all the GLFG motifs of a MAC repeat were exchanged with the NIFN motifs typical of the MIC repeats?

To address this question, hydrogels formed from a wild-type macronuclear GLFG repeat and the corresponding GLFG-NIFN mutant were challenged with human Impβ-GFP. Strikingly, the mass flux of the NTR into both the WT and mutant hydrogel was essentially identical (Figure 5.2.B). Similarly, only a small effect on the passive sieving capabilities was observed, as the partitioning coefficient of MBP-mCherry increased from ~0.4 in the WT to ~0.6 in the mutant gel (Figure 5.2.C).

These results hence indicate that the mere exchange of FG motifs is not sufficient to alter the selectivity properties of FG repeats. As a matter of fact, also the spacer regions differ significantly between the *Tetrahymena* MAC and MIC repeats (Figure 5.1). Whereas one-third of all residues in the GLFG motif-containing MAC repeats are glycines and an

---

[3] Please note that *Tetrahymena* and *Paramecium* are the only completely sequenced ciliates to date.

additional one-eight are threonines, the MIC repeats are extremely NQ-rich (>50%). In agreement with the correlation noted before (Figure 2.2.D), the MIC repeats are overall by far less hydrophobic than the MAC repeats.

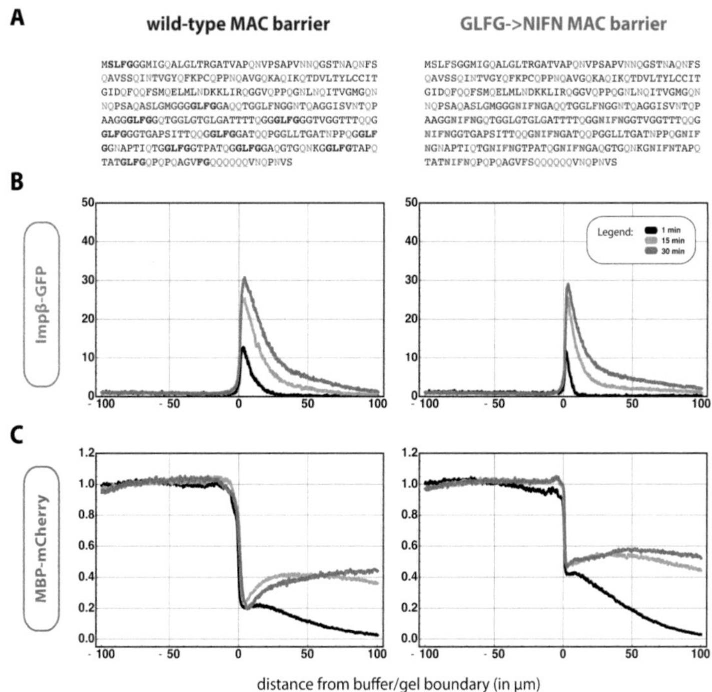

Figure 5.2. A mere exchange of the FG motifs is not sufficient to alter the specificity of the MAC hydrogels. (A) Amino acid sequences of the WT (left) and GLFG-NIFN mutated (right) TtMacNup98B repeat domains. (B) Hydrogels formed from both domains (at 200mg/ml) show the same permeability properties for GFP-tagged human Impβ (used at 1μM). The concentration profiles after the NTR influx of t = 1, 15 and 30 minutes are shown. (C) In comparison to the WT hydrogels, the partitioning coefficient of MBP-mCherry (used at 3μM) is slightly enriched from ~0.4 to ~0.6 in the mutant hydrogels.

2.4.2 The *Tetrahymena* MAC and MIC Nup98-derived FG repeat hydrogels reproduce the basic aspects of nuclear dualism.

From all known components to date (see Table 1.3), the MAC and MIC *Tetrahymena* NPC varies only in their Nup98 constituents. Hence, the MAC and MIC repeat hydrogels seem to be well suited for reconstituting nuclear dualism and to thereby gain mechanistic insight in this long-standing enigma of cell biology.

As aforementioned, nuclear dualism however not only requires specialized permeability barriers, but likewise also adapted NTRs. Here, TtImpB3 and TtImpB6 were chosen from the known *Tetrahymena* NTRs (see Table 1.3) to challenge the hypothesis that the MAC and MIC hydrogels reproduce the fundamental aspects of nucleus-specific transport. Peculiarly, TtImpB6 is the only described *Tetrahymena* NTR with a marked MIC localization (it was however also found in the MAC). TtImpB3 in contrast was exclusively localized to the MAC (Malone et al., 2008).

Due to the fact that ciliates use an alternative genetic code (see also footnote in Section 2.1.4), synthetic genes encoding for the two *Tetrahymena* NTRs were designed and synthesized to facilitate the cloning process and allow for the optimization of the respective nucleotide sequences in regard to recombinant expression in *E.coli* (see Materials and Methods). As experienced for the *Tetrahymena* FG repeats (see Section 2.1.4), also the NTRs only expressed poorly. Additionally (and unlike for the FG repeats, which were purified under denaturing conditions), the limited solubility of the bacterially expressed proteins was a problem here. The situation was partially improved (especially in regard to the expression levels) by fusing a bulky, cleavable His-MBP-tag to the N-terminus and including the fluorescent protein moiety required for tracing the NTRs in the hydrogel permeation assay (i.e. either GFP or mCherry) to the C-terminus. However, the poor solubility of TtImpB3 remained a problem, even after optimization of the expression conditions to favor folding and solubility of large exogenous proteins. Nevertheless, adequate amounts of soluble, purified TtImpB3 and TtImpB6 for challenging the MAC and MIC repeat hydrogels were eventually obtained.

Unexpectedly, when the MAC repeat hydrogels were incubated with the two *Tetrahymena* NTRs o/n, both qualitatively enriched almost equally well in the gels (Figure 5.3.A). Indeed, TtImpB6 showed a slightly higher partitioning coefficient than TtImpB3. The situation was substantially different in the MIC repeat hydrogels (Figure 5.3.B). Here, the total mass flux of TtImpB6 into the gel was by far greater than the one of TtImpB3: the observed partitioning coefficients varied by a factor of ~4 and the penetration depths approximately 2-fold. Importantly, the gels were challenged simultaneously with both NTRs in all cases, hence possible competition effects for preferred binding sites that theoretically could add to the overall selectivity should have been accounted for in the experiments.

Interestingly, when looking at the experiments focussing on the individual NTRs, it becomes evident that TtImpB3 still entered the MAC repeat gel better than the MIC repeat gel (Figure 5.3.C). The total mass of TtImpB3 taken up by the MAC gel (i.e. given by the integrals of the concentration profiles) was more than 2-fold greater than the amount taken up by the MIC gel. However, the total mass of TtImpB6 taken up by the MAC gel was much smaller than the one taken up by the MIC gel (Figure 5.3.D).

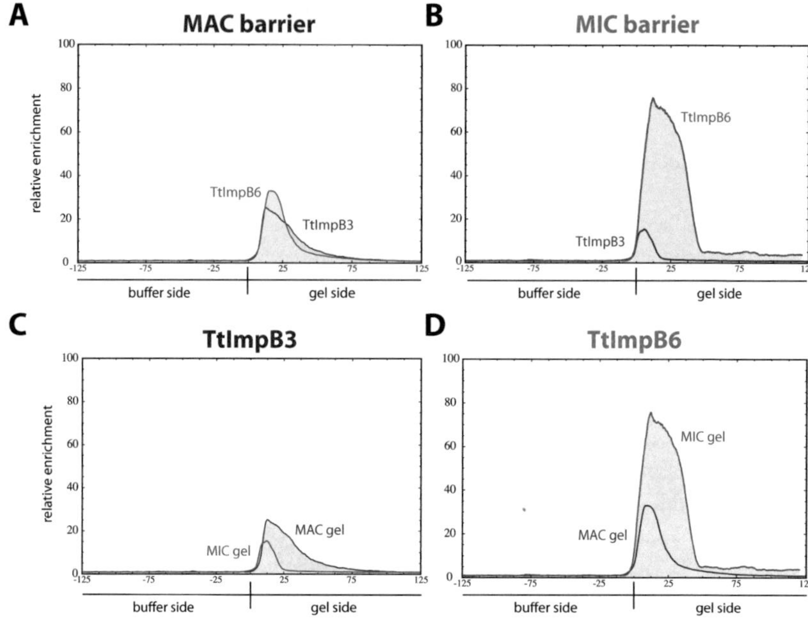

Figure 5.3. Reconstitution of nuclear dualism with the hydrogel permeation system. Hydrogels were formed at 200mg/ml from the TtMacNup98B and TtMicNup98A repeats and challenged with the MAC-specific NTR TtImpB3 or the MIC-preferring NTR TtImpB6 (Malone et al., 2008), both at 1µM. (A) Enrichment of TtImpB3 and TtImpB6 in the MAC barrier after o/n incubation at RT. The gel was simultaneously challenged with both NTRs, which were used as GFP- or mCherry fusions. Note that both NTRs enrich almost equally strong in the MAC gel. (B) Enrichment of TtImpB3 and TtImpB6 after simultaneous incubation with the MIC gel. Note that TtImpB6 enriches ~35-fold more in the MIC gel than TtImpB3. (C) The total mass of TtImpB3 taken up by the MAC gel is ~2-fold greater than the amount taken up by the MIC gel. Note that two separate gels are compared here. (D) TtImpB6 enriches ~3-fold more in the MIC compared to the MAC gel. Also here, two separate gels are compared.

Hence indeed, some of the basic aspects of nuclear dualism (as described in the literature) could be reproduced in the hydrogel system. Yet, true orthogonality of the reconstituted transport apparatuses was certainly not observed, despite having used the most specialized NTRs and the most divergent FG repeats known in *Tetrahymena* to date. In such a case, one would expect the flux of a given NTR into the two gels to vary by a factor of at least 10 (meaning that 90% of the NTR species enters the one, but not the other gel).

2.4.3 Biochemical identification of nucleus-specific NTRs in *Tetrahymena*.

Still, the possibility remains that not all, and in particular the most crucial, *Tetrahymena* NTRs have been identified yet via the *in silico* approach followed by (Malone et al., 2008). It should however be possible to use the macronuclear GLFG and micronuclear IF repeats as affinity baits to bind NTRs from a *Tetrahymena* extract and thereby identify previously unknown factors with the help of mass spectrometry. Therefore, a *Tetrahymena* cell culture and protocols for extract preparation were first established in the lab.

Growing *Tetrahymena* is fairly easy. Indeed, this organisms was originally classified belonging to the 'Infusoria', an ancient supergroup of 19. century systematics encompassing ciliates, euglenoids and other protozoa that commonly thrive on classical hay infusions. Today, rich media based on proteose peptone and yeast extract are available that allow growing *Tetrahymena* to densities of up to two million cells/ml (Kiy and Tiedtke, 1992; Hellenbroich et al., 1999). In this study, *Tetrahymena* was routinely grown in a 13l bench-top fermenter (see Materials and Methods).

Extract preparation however proved to be much more problematic due to the levels of protease activity resulting from both the proteases secreted to the medium and the proteases released upon cell lysis. As shown in Figure 5.4.A, a reporter protein was hardly stable in extracts prepared with the addition of standard protease inhibitor cocktails (here: Roche Complete protease inhibitor tablets or Sigma-Aldrich protease inhibitor cocktail). Moreover, the smeared lanes on the gel further illustrate that also endogenous proteins suffer from proteolytic degradation. To solve these problems, the effect of various protease inhibitors was systematically tested. In agreement with the finding that *Tetrahymena* contains unusual high amounts of cysteine proteases (Eisen et al., 2006), the cysteine protease inhibitor E64 (in combination with standard serine protease inhibitors such as PMSF and metalloprotease inhibitors such as EDTA) greatly improved the situation (once titrated to an optimum working concentration). Now, the same reporter protein remains stable even after two hour incubation in the extracts at 30°C, the optimum growth temperature of *Tetrahymena* (Figure 5.4.A). Importantly, E64 attacks and covalently modifies the active site cysteine residue and hence is a very potent inhibitor of cysteine proteases. Indeed, non-covalent cysteine protease inhibitors such as Leupeptin were found to be rather ineffective.

Taken together, high quality whole cell *Tetrahymena* extracts were obtained after optimization (see Figure 5.4.B).

To next identify NTRs in these extracts, the MAC and MIC repeats were coupled to Sepharose beads (see Methods) and used as baits in classical pull-down experiments. As controls, low substitution phenyl-Sepharose, which had previously been shown to efficiently bind NTRs (Ribbeck and Görlich, 2002), and mock treated Sepharose to visualize the background binding have been included.

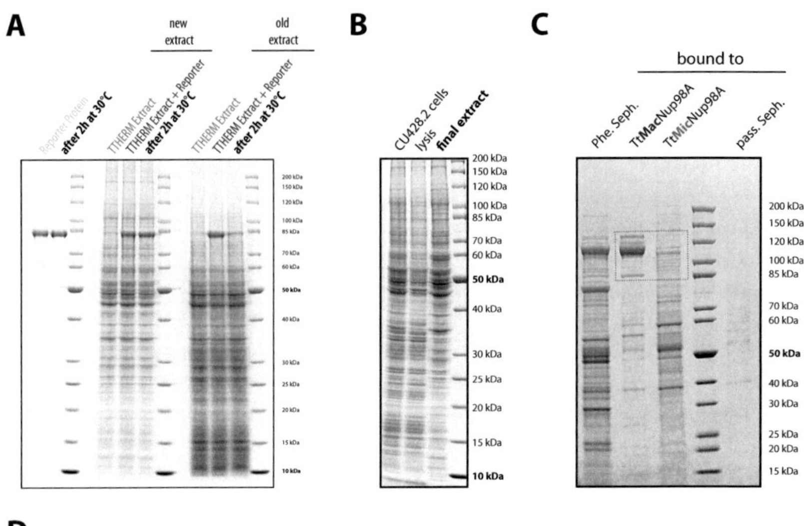

MS/MS Analysis of the Pulldowns on the TtNup98 Orthologs

	MAC			MIC			
	Experiment			Experiment			
	(1) full lane	(2) single band	(3) single band	(1) full lane	(2) single band	(3) single band	
TtImpB12	41		5	0		0	
TtCrm1	37		5	0		0	
TtImpB14		14			0		
TtImpB5	7			0			MAC only
TtImpB7	6			0			
Gle2/Rae1	6			0			
TtImpA3	3			0			
TtImpA14	2			0			
TtImpB3	320	136	47	2	0	6	
TtImpB2	160	69	23	1	0	2	
TtImpB13	107	3		5	0		
RRP12	102	1		3	0		strong MAC preference
TtImpB4	90	77	9	1	0	0	
TtImpB11	65	44	16	1	0	0	
TtImpA1	29			2			
TtImpB1	79			32			
Trn3	47	5		13	0		MAC preference
Trn3b	9			3			
TtImpB8	28	0		24	3		
EF hand	42	1	9	78	0	18	
TtImpB6	16	23	2	128	83	12	MIC preference
TtCAS	5			47			
TtRan	4			11			

Figure 5.4. Biochemical identification of nucleus-specific *Tetrahymena* NTRs. (A) 4µM of an NLS-Sp68-MBP-mCherry fusion were used as a proteolysis-susceptible substrate and incubated with whole cell *Tetrahymena* extracts for 2 hours at 30°C to assay protease contamination. Note that prior to optimization of the protease inhibitor mix, the extracts were heavily contaminated with proteases, as can be seen by (i) the complete degradation of the reporter and (ii) the smeared lanes on the SDS-gel. (B) Quality control of the final *Tetrahymena* extracts used for experimentation after optimization. (C) TtMacNup98A and TtMicNup98A were covalently linked to a Sepharose matrix and used as baits to pull-down NTRs from

Tetrahymena extracts. The baits and extracts were incubated for 1-2h at 4°C (see Methods). As a positive control for NTR binding, phenyl sepharose was used (Ribbeck and Görlich, 2002). The dashed box indicates the molecular weight range in which most NTRs have been identified by subsequent mass spectrometry analysis. (D) Summary of the mass spectrometry results. The numbers of assigned peptides for three independent experiments are given. Highlighted in red are NTRs newly identified in the present study. The NTRs are grouped according to their preference for either type of repeat, which was estimated based on the ratio of the assigned spectra found for a given hit in the MAC or MIC pull-downs. The results were in good agreement with previously reported localization data for the subset of known *Tetrahymena* NTRs (Malone et al., 2008). Note that no MIC-specific NTRs were identified.

As shown in Figure 5.4.C, hardly any unspecific binding of proteins in the size range of known NTRs was observed, whereas the most prominent bands pulled-down by the MAC repeats were found in this window (see gray box). Subsequent analysis by mass spectrometry verified the presence of NTRs here. In total, five NTRs (i.e. four putative Imps and one putative Exp), two putative Impα-like import adaptors and the RNA export factor Gle2/Rae1 were indeed found exclusively in the pull-downs on the macronuclear GLFG repeat (see Figure 5.4.D). Notably, two of the four Impβs and one of the two putative Impαs, as well as the *Tetrahymena* Gle2/Rae1 homolog, have not been described before. Additionally, six other putative Imps and the RNA export factor RRP12 have been predominantly pulled-down by the MAC repeat.

Unexpectedly though, no MIC-specific NTR was identified. Rather, five putative Imps (amongst them the here newly discovered *Tetrahymena* Trn3 and Trn3b homologs) and one putative Exp have been found to also bind to the MIC repeats. Interestingly, also the *Tetrahymena* ortholog of Impβ was amongst them. The most prominent protein pulled-down by TtMicNup98A however was TtImpB6. Another prominent hit was an unknown protein that bears partial homology to the ARM repeats of Imp-αs and contain terminal EF hand Ca^{2+} binding motifs. Interestingly, also the previously uncharacterized *Tetrahymena* homolog of CAS, the exportin required to recycle Impαs in the classical nuclear import pathway (Kutay et al., 1997a), was predominantly found in the MIC repeat pull-downs.

In summary, four novel proteins implicated with nuclear transport were exclusively found in the pull-downs on the MAC repeats, whereas no such hit was found in the MIC repeat-purified fraction. Please note that the results have been qualitatively reproduced multiple times, even when varying parameters such as the lysate-to-matrix ratio or using extracts enriched in NTRs via ammonium sulfate precipitation. Certainly, the newly identified NTRs are interesting candidates to test in the hydrogel system.

2.4.4 Import adaptors may aid in nucleus-specific protein transport.

Interestingly however, out of the eleven *Tetrahymena* Imp-α family members for which localization data is available, nine were found to be MIC-specific (Malone et al., 2008). What is hence the role of these import adaptors in nuclear dualism?

Given the structural resemblance of the HEAT repeats characteristic of Impβ-like NTRs and the ARM repeats of Imp-αs, it might be very well possible that the latter evolved to become more NTR-like in *Tetrahymena*, hence potentially aiding in mediating nucleus-specific nucleocytoplasmic transport. Notably, the hydrogel reconstitution system described above is ideally suited to test this hypothesis. TtIMA1, as well as TtIMA8 and TtIMA12 were chosen for experimentation, primarily because they show the most striking MAC or MIC localization during vegetative growth, respectively (Malone et al., 2008). Please note that TtIMA1 moreover bears the strongest resemblance to canonical Imp-αs such as yeast Srp1. Unlike the *Tetrahymena* Imp-βs, all tested Imp-αs expressed fine in *E.coli* and behaved well during protein purification.

As expected, the *Tetrahymena* Imp-αs behaved indeed neither in the MAC, nor MIC repeat hydrogels as truly inert molecules (Figures 5.5.A and 5.5.B). TtIMA1 was arguably an exception, as it only showed a partitioning coefficient slightly above the one observed for mCherry (~0.75 and ~0.6, respectively) after o/n incubation with the MAC gel (Figure 5.5.A top and bottom panel, respectively). In contrast, TtIMA8 equilibrated between the buffer and gel phase, and TtIMA12 even slightly enriched in the MAC gels.

Strikingly however, all TtIMAs were 'soaked up' by the MIC repeat hydrogel in great amounts (Figure 5.5.B). For example, TtIMA8 penetrated up to ~75µm into the MIC gel and stunningly showed an ~175-times greater partitioning coefficient in the MIC gel compared to the MAC gel. Notably, in similar experiments where MIC gels were incubated with TtImpB6, the observed partitioning coefficient reached only ~80 and the penetration depth was limited to ~50µm (Figure 5.3).

These findings further nourish earlier speculations suggesting that the subgroup of MIC-specific Imp-αs may contribute significantly to the overall nuclear specificity of a transport species (Malone et al., 2008) or even function as specialized NTRs. However, also the MAC-specific TtIMA1 and mCherry enriched in the MIC gel compared to the MAC gel. Whereas for the latter, the partitioning coefficient was only increased ~4-fold, the difference amounts to a factor of ~65 in the former case, which argues for a rather specific interaction. Note though that the partitioning coefficient of TtIMA1 in the MIC gel is only ~1/3 as high as the one of MIC-specific TtIMA8 in the same gel.

Additional evidence for the ability of all three TtIMAs to specifically bind FG repeats comes from a classical binding assay (Figure 5.5.C): in contrast to Srp1, the sole yeast nuclear import adaptor, the *Tetrahymena* Imp-αs bound to the immobilized MAC and MIC repeats, as well as to the regular Nsp1 FxFG repeats. Remarkably, the *Tetrahymena* import adaptors bound the *Tetrahymena* repeats even better than the *bona fide* human NTR Impβ. This was especially true for the MIC repeats, but not for the Nsp1 repeats.

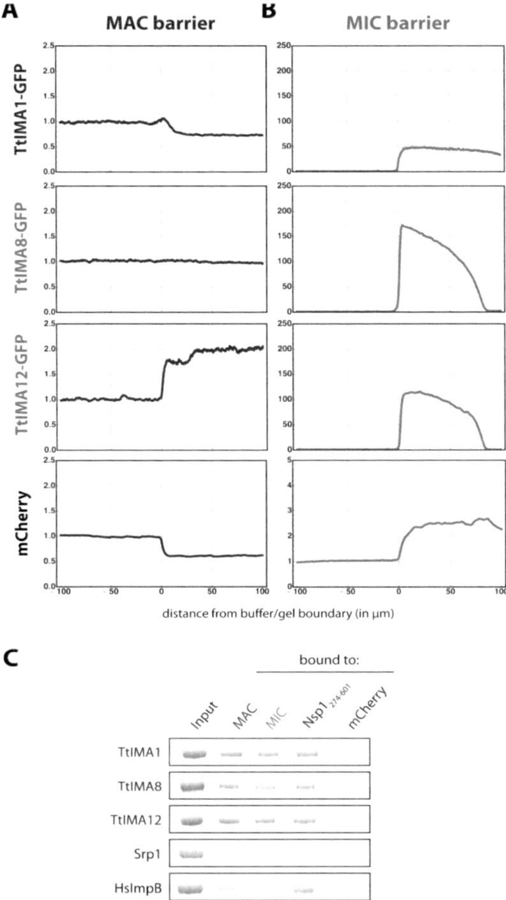

Figure 5.5. Selected *Tetrahymena* import adaptors especially enrich in MIC hydrogels. TtMacNup98B (A) and TtMicNup98A (B) hydrogels (both at 200mg/ml) were challenged o/n at RT with 1µM of the indicated GFP-labelled import adaptor or 3µM of a mCherry control. Please note the different scales. (A) With partitioning coefficients ranging from ~0.75 to ~2.0, the *Tetrahymena* proteins did not behave as truly inert molecules in the MAC gels. (B) Even more surprising, the import adaptors enriched up to ~175-fold in the MIC gel. Note however that the MIC gel also absorbed mCherry, albeit to a much lesser extend. (C) A classical binding assay confirmed that the *Tetrahymena* import adaptors, in contrast to the sole yeast Impα Srp1, specifically bound FG repeats. Note that the canonical human NTR Impβ bound the MAC and MIC repeats much worse than the *Tetrahymena* Impα homologs. For the experiments, 30µg of the indicated baits were immobilized to Ni^{2+} 500Å silica beads and incubated with 2µM of the preys for 1 hour at RT. The binding experiments were analyzed with SDS-PAGE.

2.4.5 TtCAS binds TtImpαs in a RanGTP-dependent manner and enriches in both MAC and MIC repeat hydrogels.

Notably, another key conceptual difference apart from productive FG repeat binding additionally sets classical Imp-αs apart from *bona fide* NTRs: they require the help of CAS to ultimately release their bound cargo in the nucleus and return to the cytoplasm (Kutay et al., 1997a). However, it is unlikely that TtIMAs also evolved independent of this regulation, even when considering the above postulated hypothesis that at least some TtIMAs might have taken over as nucleus-specific NTRs independent of the help by Imp-β family members. The main argument in this regard is that TtIMAs would have to become directly regulated by Ran in order to control cargo release and bestow directionality to the transport process, which, given the large-scale concerted conformational dynamics required in Impβ-like NTRs to achieve this, is evidently much more complex than to simply surface-optimize an already favorably designed molecule to obtain novel IF repeat binding sites.

Note that nevertheless, an import cycle where only an Impα-family member and TtCAS were required, but no additional Impβ-like NTR, would still be energetically less expensive than the classical import pathway (in terms of hydrolyzed GTP molecules per complete cycle). In such a situation, the RanGTP required to separate the α- and β-importins would now be obsolete, leaving the formation of the trimeric export complex with TtCAS the only point where metabolic energy is fed into the system.

This reasoning was experimentally first surveyed by testing whether the *Tetrahymena* CAS homolog identified in this study (see Section 2.4.3) can still interact with TtIMAs showing NTR-like behavior in the hydrogel permeation assay. The three available import adaptors were expressed as fusions with the IgG-binding domain from bacterial protein A (the so called ZZ-tag), immobilized on IgG-Sepharose and incubated with recombinantly expressed[4] TtCAS in a buffered bacterial lysate. As expected, the exportin was indeed recovered with TtIMA1 and TtIMA8, but only if the reaction mixtures were supplemented with TtRanQ70L (Figure 5.6.A). This mutant of Ran is unable to hydrolyze GTP and thus locks the molecule in its GTP-bound form (Klebe et al., 1995; Milles and Lemke, 2011). Note however that TtIMA12 was incompetent of binding TtCAS.

The finding that TtCAS binds to both MAC- and MIC-localized TtIMAs suggests that it has to pass both the MAC and MIC permeability barrier to participate in productive transport cycles as outlined above. In agreement with this notion, TtCAS was pulled-out of whole cell *Tetrahymena* extracts by the MAC and MIC repeats alike (see Figure 5.4). As hence expected, TtCAS (in a fusion with GFP) also entered both MAC and MIC repeat hydrogels (Figure 5.6.B). However, while the exportin was rather stuck in the MIC gel (diffusing only ~10µm far in 30 minutes), it penetrated deep into the MAC gel within the

---

[4] Note that TtCAS was the only Impβ-like NTR from *Tetrahymena* studied here that on its own expressed well and was sufficiently soluble.

same time span. Note though that the actual NPC permeability barrier is only in the range of ~50nm thick.

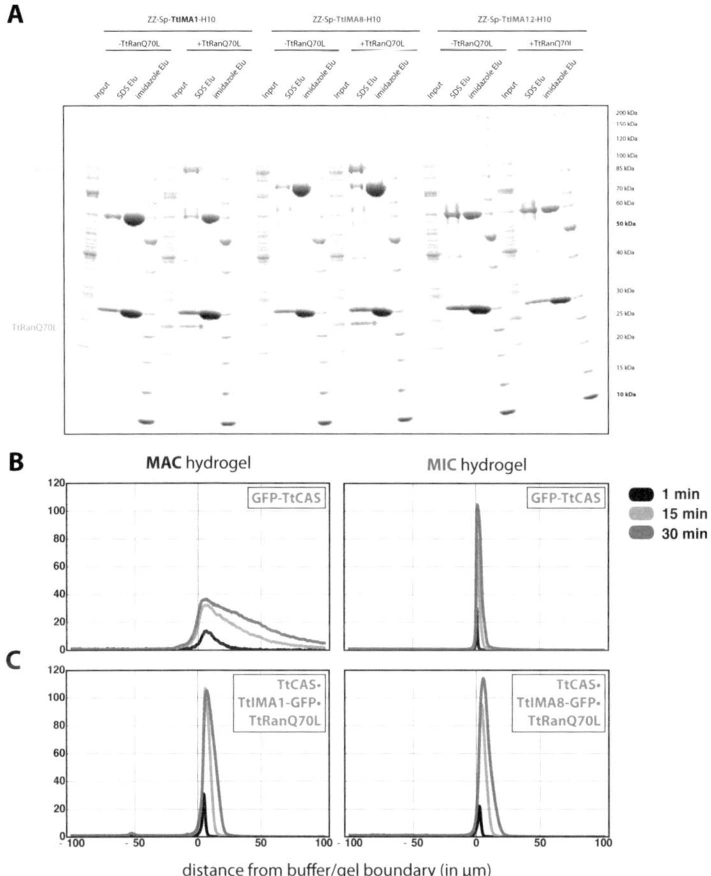

Figure 5.6. Characterization of the *Tetrahymena* CAS homolog. (A) 1.5µM of the indicated ZZ-tagged *Tetrahymena* import adaptors were incubated with 2µM of the *Tetrahymena* CAS homolog in a buffered bacterial lysate supplemented with 3µM TtRanQ70L for 1 hour at 4°C. In control binding experiments, the RanGTP mimic was left out. Note that only TtIMA1 and TtIMA8 bound TtCAS in a RanGTP-dependent manner. Next, TtMacNup98B and TtMicNup98A hydrogels (200mg/ml) were incubated with GFP-labeled TtCAS (B) or the identified TtCAS•import adaptor•TtRanQ70L complexes, as denoted (C). Influx was followed over time. Depicted here are the concentration profiles of the indicated species after t = 1, 15 and 30 minutes. As TtIMA1 was reported to be MAC- and TtIMA8 to be MIC-specific (Malone et al., 2008), two different nucleus-specific species were presumably available for experimentation. Note that the partitioning coefficient of the trimeric complex in the MAC gel was much higher than for TtCAS alone. However, the diffusion depth was greatly reduced. In contrast, the intra-gel diffusion of the trimeric complex in the MIC gel was facilitated compared to the unladen NTR.

Interestingly, a MIC-specific trimeric complex comprising TtCAS, TtIMA8 and TtRanQ70L diffused much better within the MIC hydrogel (penetrating approximately twice as far into the gel), despite its overall larger size (Figure 5.6.C, right panel). Analogously, also the MAC-specific TtCAS, TtIMA1 and TtRanQ70L complex ingressed well into the MAC hydrogel (Figure 5.6.C), although it reached not nearly as far into the gel as the NTR alone. However, the partitioning coefficient of the trimeric complex was much larger than the one observed for TtCAS alone.

Taken together, the evidence provided here suggests that TtCAS can (i) pass through both the MAC and MIC permeability barriers, (ii) bind TtIMAs in a RanGTP-dependent manner in either compartment and (iii) mediate the translocation of the TtIMAs back to the cytoplasm, thereby completing MAC- or MIC-localized transport cycles alike. Yet, the question remains whether all TtIMAs require the help of Impβ-like NTRs in all instances of ciliate nuclear import.

2.4.6 The investigated TtIMAs contain putative IBBs and bind to the *Tetrahymena* Impβ homolog.

Normally, Impβ recognizes the Importin β-binding (IBB) domain of Impα, a high affinity nuclear localization sequence (NLS) rich in arginine and lysine residues (Görlich et al., 1996a; Izaurralde et al., 1997; Görlich et al., 2003). Hence, do the *Tetrahymena* Impα-family members studied here possess similar basic patches?

Indeed, a thorough sequence analysis guided by the initial mapping of the IBB domain (Görlich et al., 1996a) and the crystal structure of the IBB domain bound to Impβ (Cingolani et al., 1999) revealed putative IBBs in the N-termini of TtIMA1, TtIMA8 and TtIMA12 (Figure 5.7.A).

To determine whether the extracted sequences can be bound by NTRs and, if so, by which, pull-down experiments with immobilized full length TtIMAs or their presumed IBBs were performed and analyzed by mass spectrometry. Due to disambiguates in regard to the domain boundaries in the case of TtIMA8, two versions varying slightly in their total length were tested. For the experiments, the proteins or peptides were eventually expressed as ZZ tag-fusions and tethered to IgG-Sepharose (see Methods). *Tetrahymena* extracts were added and incubated together with the baits for 4h at 4°C (see also Methods).

Notably, a single 85kDa band subsequently identified as the *Tetrahymena* Impβ homolog (referred to as TtImpB1 here and in the literature) was pulled-out by all putative IBBs (Figure 5.7.B). In contrast, all tested truncations lacking the putative IBBs did not bind TtImpB1.

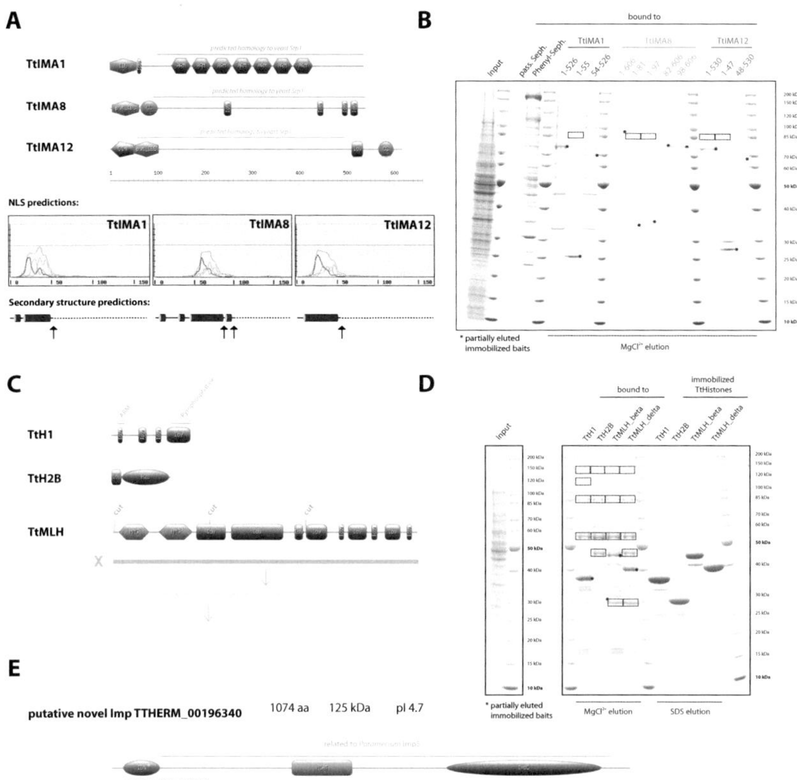

Figure 5.7. Biochemical identification of nucleus-specific transport species. (A) Domain overview of the studied *Tetrahymena* import adaptors. IBB and put_IBB denote verified and putative Importinβ-binding domains (Malone et al., 2008), ARM indicates the presence ARM repeats, CC and low represent coiled-coil or low complexity regions, respectively. Note that TtIMA1 shows the greatest resemblance to known Impβ-like import adaptors. To define the domain boundaries of the IBBs, the amino acid sequence (guided by Görlich et al., 1996a and Cingolani et al., 1999), NLS (Nguyen Ba et al., 2009) and secondary structure (Cole et al., 2008) predictions were employed. Indicated by the arrows are the domain boundaries chosen for the pull-downs. In case of TtIMA8, two different constructs were tested. (B) Pull-downs with the putative IBB domains from TtIMA1, TtIMA8 and TtIMA12 to identify nucleus-specific transport complexes. 60µg of the indicated baits were immobilized to IgG-Sepharose and incubated with whole-cell Tetrahymena extracts for 1-2 hours at 4°C. As negative and positive controls, ZZ-tag passivated IgG-Sepharose and phenyl sepharose were used, respectively. The bound material was eluted with 1M MgCl2. Subsequent analysis of the pull-downs with mass spectrometry revealed that the only NTR bound to the putative IBBs was in all cases TtImpB1, which runs at 85kDa (as indicated by the dashed boxes). Asterisks denote partially eluted baits. (C) Domain organization of the major macronulcear (TtH1) and micronulcear (MLH) linker histones, as well as the omnipresent histone H2B. Note that the unusual Tetrahymena MLH is post-translationally processed into three functional variants. As it is still debated whether the proteolytic processing occurs before or after nuclear import, constructs for the recombinant expression of both the precursor protein and processed products have been designed. However, only the β and δ variants could be purified and used for further experimentation so far. (D) Pull-downs with the indicated *Tetrahymena* histone species, which are

nucleus-specific cargoes. Again, TtImpB1 was found as a prominent hit in all pull-downs (indicated by the boxed 85kDa-band). Likewise, a 125kDa previously uncharacterized protein with partial homology to the known histone importer Imp5 and a 55kDa putative histone chaperone were found to interact with all histones. Specifically bound to the macronuclear TtH1 histone was the 115kDa known MAC-specific NTR TtImpB3, which also bears homology to Imp5. A 48kDa homolog of the nucleosome assembly protein (NAP) was found to interact with TtH2B and TtMLH_δ. Various 14-3-3 proteins inbetween 25-30kDa bound to the two TtMLH variants. (E) Domain overview of the newly identified putative Imp5-family member. IBN denotes a stretch with homology to the Importinβ N-terminus, HEAT represents putative HEAT repeats and Imp5 a conserved part of the sequence with homology to Imp5. Overall, the majority of the protein bears close resemblance to the *Paramecium* Imp5 homolog.

Interestingly, the TtIMA1-ΔIBB body also bound a plethora of other proteins (not analyzed so far), possibly reflecting cargoes. TtIMA8 and TtIMA12 however seem to recognize by far less presumed cargoes, as hardly any bands were seen in the respective ΔIBB-adaptor pull-downs (Figure 5.7.B).

Of the full-length TtIMAs, only TtIMA12 also bound TtImpB1. This finding was unexpected, as in its cargo-free state, Impα normally binds its own IBB via its NLS recognition site, thus residing in an auto-inhibited state (enforced by CAS) ensuring that a just imported cargo is not recycled back to the cytoplasm (Kobe, 1999).

In summary, given that TtImpB1 localizes to both *Tetrahymena* nuclei (Malone et al., 2008) and was identified in both the MAC and MIC repeat pull-down fractions (Figure 5.4), a picture emerges in which nucleus-specific import adaptors recognize exclusive cargoes and modify the nuclear specificity of TtImpB1. Challenging this hypothesis experimentally using the hydrogel reconstitution system established here however remains to be done, mainly because TtImpB1 so far could not be expressed and purified to the required quantity or quality.

2.4.7 Identification of a putative novel Imp5-like NTR suggests that *Tetrahymena* contains multiple Imp5 homologs possibly involved in nucleus-specific protein import.

Amendatory to identifying nucleus-specific transport species by fractionation of whole cell *Tetrahymena* extracts with MAC- or MIC-derived FG repeats, known nucleus-specific cargoes were alternatively used as affinity baits in complementary pull-down experiments. In particular, the MAC-specific canonical H1 linker histone and the micronuclear linker histones (MLH) were chosen, as they are arguably the best described examples of such proteins (see e.g. Johmann and Gorovsky, 1976; Gorovsky et al., 1978; Wu et al., 1994) and commonly used as discrete markers in localization studies (e.g White et al., 1989; Iwamoto et al., 2009). As discussed in the Introduction, the MLHs are first expressed as a fusion protein and then proteolytically processed into the three linker peptides (see Figure 5.7.C). However, it is not yet clear whether the cleavages occur immediately after translation in the cytoplasm or only following nuclear import, especially since both intermediate and final products have been found in both compartments alike (Allis et al., 1984; Wu et al., 1994). Hence, the synthetic genes for the recombinant expression of the full-length MLH precursor, as well as the individual parts, were designed and ordered (see Materials). Unfortunately however, only the β and δ linker histones could be expressed and purified up to this point. The pull-down experiments were in turn only performed with a subset of all possible import substrates. Additionally, the core histone H2B, a component of both the MAC and MIC chromatin (Gorovsky et al., 1978), was included as well.

Analysis of the pull-downs with mass spectrometry revealed that three proteins were bound by all studied histones (Figure 5.7.D): (i) the above mentioned TtImpB1, (ii) a 55kDa putative histone chaperone with predicted peptidyl prolyl cis-trans isomerase activity and (iii) a previously uncharacterized 125kDa protein (Figure 5.7.E) whose closest homolog is an unknown *Paramecium* protein resembling Imp5 (Deane et al., 1997; Jäkel and Görlich, 1998), a NTR recognized to be involved in histone import (Bayliss et al., 2000; Baake et al., 2001; Mosammaparast et al., 2001; Mühlhäusser et al., 2001; Mosammaparast et al., 2002). Notably, this newly identified *Tetrahymena* importin shows an acidic pI like NTRs in general and contains a prominent acidic stretch in its Imp5-like region. Additionally, the MAC-localized TtH1 bound the MAC-specific TtImpB3. Histone δ and TtH2B in turn selectively pulled-out the putative *Tetrahymena* nucleosome assembly protein (NAP) homolog. The two MLH variants moreover interacted with various 14-3-3 proteins.

Taken together, these preliminary results implicate TtImpB1 and TtImpB3 in the import pathway of TtH1 into the MAC and suggest that the two NTRs might interact to form a MAC-specific transport species together with their histone cargo. In further support of this notion, TtImpB3 also shows homology to Imp5. Together with the identification of the novel Imp5-like NTR described above, which additionally might aide in histone import (probably both into the MAC and MIC), this possibly hints at a situation were the nuclear specificity of TtImpB1 is not only modified by different Impα-like import adaptors, but (at least in the case of histone import) also by adapted Imp5-like NTRs. Importantly however, it still remains to be tested whether these factors indeed interact as hypothesized here and how they behave in MAC and MIC repeat hydrogels.

2.4.8 Biochemical identification of additional barrier-constituting *Tetrahymena* Nups.

Based on the finding that MAC and MIC repeat hydrogels only reproduced basic aspects of nuclear dualism (see Figure 5.3.), the idea that multiple transport-associated factors interact to form nucleus-specific import complexes was extensively addressed up to this point in order to completely reconstitute this phenomenon.

However, recall that (a) the MAC repeat hydrogels were rather tight barriers, which (compared to other FG hydrogels) only sustain relatively small NTR fluxes, and (b) the MIC repeat hydrogels showed significant background binding of supposedly inert molecules such as MBP-mCherry. As shown in Section 2.2, the regular $Nsp1_{274-601}$ FSFG repeats could however alleviate the latter problem. Hence, it is highly likely that other *Tetrahymena* Nups are required to fulfill the task of loosening or passivating the barrier of the MAC or MIC pore, respectively. Additionally, it is by no means unfounded to assume that such Nups also significantly contribute to the overall selectivity of the two barriers. Yet, only few FG Nups are known in *Tetrahymena* so far (see Table 1.3).

A possible approach to biochemically identify additional FG repeats in *Tetrahymena* is to use natural binding partners as baits in pull-down experiments. NTRs immediately come to mind. However, these usually only posses low affinity for FG repeats (Ribbeck and Görlich, 2001; Tetenbaum-Novatt et al., 2012). An exception is the $Imp\beta_{45-462}$ fragment, which indeed binds FG repeats so strongly that it blocks multiple nuclear transport pathwys (Kutay et al., 1997b). This mutant was further optimized for pull-down experiments by replacing an acidic loop in its C-terminal cargo-recognition domain with a hydrophilic spacer in order to prevent unspecific background binding as a result of electrostatic interactions (T. Pleiner, personal communication). Additionally, it was genetically fused to an Avi-tag, to which a biotin moiety can be attached by the bacterial biotin ligase BirA, which was co-expressed with the Impβ mutant. This trick allowed immobilization of the protein to Streptavidin-Agarose. The strategy to pull out and identify FG repeats from *Tetrahymena* extracts using the optimized $Imp\beta_{45-462}$ mutant is summarized in Figure 5.8. Please note that the purified biotinylated $Imp\beta_{45-462}$ mutant used in this study was a kind gift of T. Pleiner.

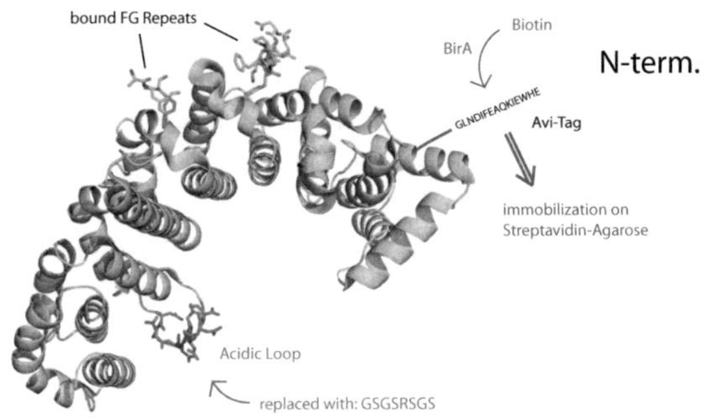

Figure 5.8. Strategy for the biochemical identification of additional FG repeats in *Tetrahymena thermophila*. Highlighted here in the Impβ structure solved by (Bayliss et al., 2000; 2002b) are the features of the $Imp\beta_{45-462}$ fragment designed by T. Pleiner for the depletion of FG repeats from cell extracts. Fused to the N-terminus is an Avi-tag, to which a biotin moiety was added by a co-expressed, engineered *E.coli* BirA biotin ligase. The biotinylated $Imp\beta_{45-462}$ mutant can hence be immobilized to Streptavidin-Agarose, thereby creating affinity matrices for the use with extracts. Note that the acidic loop in the C-terminus of the protein has been removed with a flexible spacer to reduce unspecific background binding due to electrostatic interactions. Indicated are also the primary FG repeat-binding sites between the HEAT repeats of Impβ. The biotinylated $Imp\beta_{45-462}$ used in the present study was a kind gift from T. Pleiner.

Importantly however, such a well-characterized human NTR mutant promiscuously binding FG repeats is on the one hand certainly a highly accessible, easy-to-use and powerful tool, yet on the other hand, it clearly introduces an experimental bias, as out of all *Tetrahymena* FG repeats one will select for only such species that can bind this particular NTR, without knowing what else has been missed. Hence in the following, no complete inventory of all involved factors can be provided. Rather, only putative additional FG repeats might be unraveled, as mentioned above.

As expected based on the finding that full-length human Impβ can enter MAC repeat hydrogels (Figure 2.7), also the Impβ$_{45-462}$ mutant was able to bind both TtMacNup98A and TtMacNup98B (Figure 4.9.A). From the currently known *Tetrahymena* FG repeats (Table 1.3), it also pulled-out TtNup308. Missing however are the two remaining reported FG Nups, i.e. TtNup50 and TtNup54. In agreement with the results of the binding assay shown in Figure 5.6.C, the Impβ mutant (like the full length human Impβ) did not bind the MIC-specific Nup98 homologs.

Strikingly however, it was able to interact with two uncharacterized proteins lacking *bona fide* FG motifs, but containing alternative hydrophobic patches in their predicted non-globular portions. These regions moreover show the amino acid bias typical for FG repeats (Figure 5.9.C). Taken together, the two proteins were termed TtNup72 and TtNup127 based on their respective predicted molecular weight (as conventionally done in the field). To test whether these unconventional repeats can form hydrogel-based barriers, the phenylalanine- and methionine-rich non-globular portion of TtNup72 (Figure 5.9.D) was purified from *E.coli* following the same protocol as for canonical FG repeats (see Methods). As expected, the protein indeed formed homogenous hydrogels that moreover allowed the enrichment and intra-gel diffusion of NTRs, whilst remaining a firm barrier for GFP-sized inert molecules (Figure 5.9.E).

Additionally, three additional putative new FG repeats featuring *bona fide* FG motifs were found and termed TtNup114, TtNup117 and TtNup353 (Figure 5.9.B). Strikingly, despite resembling other NQ-rich FG repeats in its primary amino acid sequence, the predicted FG repeat region of TtNup117 has a net charge of −12, which is atypical for the normally positively charged FG repeats. To the contrary, the putative TtNup353-derived FG repeats contain an overall charge of +20 (Figure 5.9.C).

In summary, despite the fact that a number of proteins specifically interacting with human Impβ and withal containing non-globular regions resembling known FG repeats in their amino acid composition were found here, their subcellular localization in *Tetrahymena* needs to be studied next to confirm their classification as Nups and asses their potential contribution to the permeability barriers of the two *Tetrahymena* nuclei.

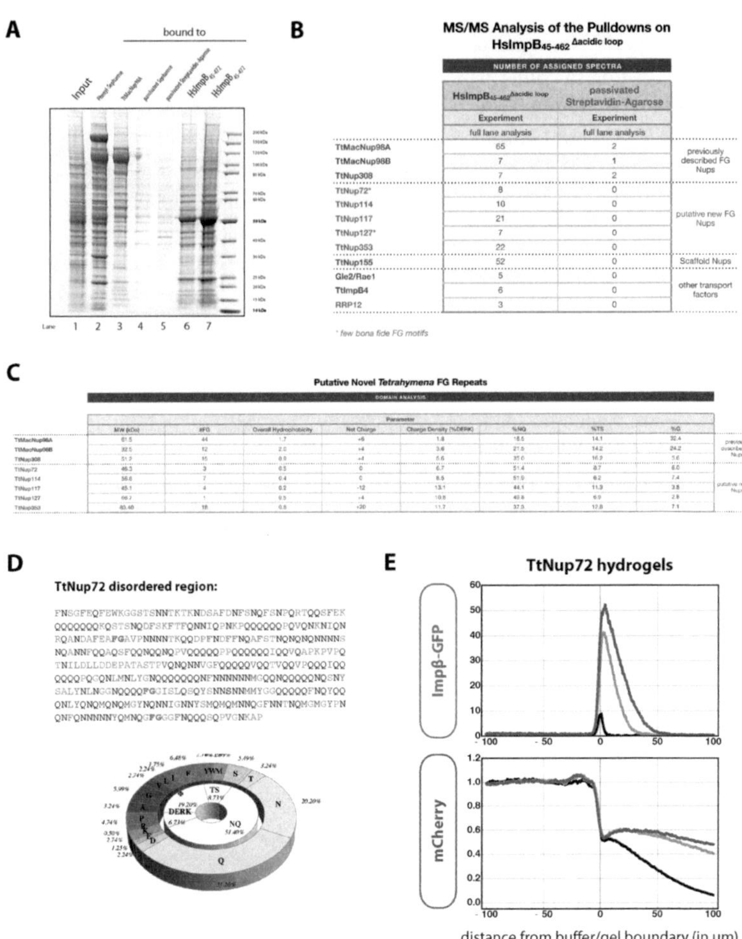

Figure 5.9. Biochemical identification of additional FG repeats in *Tetrahymena*. (A) 130µg biotinylated Impβ45-462 was bound to Streptavidin-agarose and incubated with whole-cell *Tetrahymena* extracts for 1 hour at 4°C. As a negative control, biotin-passivated Streptavidin-agarose was included. For comparison and quality control of the extract, phenyl-sepharose and the macronuclear GLFG repeats from TtMacNup98A were also used as baits to visualize NTR binding. Note that different amounts of the Impβ45-462 pull-downs were loaded in lanes six and seven. Bound proteins were identified with mass spectrometry, as described in the Methods. (B) Summary of the mass spectrometry analysis. Shown are the numbers of uniquely assigned spectra for known and putative new *Tetrahymena* FG Nups in the Impβ45-462 or control pull-downs. (C) Domain analysis of the identified FG Nups. See main text for discussion. (D) Amino acid sequence and composition of the non-globular, NQ-rich domain from putative TtNup72. (E) The disordered domain from TtNup72 forms hydrogels that can be well penetrated by human Impβ-GFP, but hardly by mCherry. The hydrogels were formed at 200mg/ml and challenged with 1µM of the active and 3µM of the passive species. Shown here are the concentration profiles after t = 1, 15 and 30 minutes of influx.

3 Discussion

NPCs are a synapomorphous feature of all Eukaryotes, providing exclusive gateways into and out of the nucleus and hence crucially participating in controlling the spatiotemporal distribution of macromolecules in nucleated cells. The functional hallmark of these gigantic protein assemblies is a permeability barrier, which effectively prevents the passive diffusion of inert objects exceeding ~5nm (Mohr et al., 2009), but paradoxically nevertheless allows the facilitated transport of even larger molecules, given that they are chaperoned by NTRs (Görlich and Kutay, 1999; Fried and Kutay, 2003).

As discussed in the Introduction, the NPC permeability barrier is formed by non-globular FG repeats. Importantly, NTRs bind to FG repeats during facilitated transport (Bayliss et al., 1999). However, simple binary interactions between NTRs and FG repeats can neither explain the observed passage rates for translocating species, nor provide a plausible explanation for the size-dependent sieving effect of the NPC permeability barrier. The selective phase model hence suggests that FG repeats also interact with each other, in turn forming a three-dimensional sieve-like protein meshwork that can be locally, transiently and specifically opened by NTRs to allow the passage of translocating species above the mesh size without disturbing the overall barrier integrity.

However, the ability of FG repeats to interact is highly disputed in the nuclear transport field. In fact, it is the crucial point in which most models for nucleocytoplasmic transport differ (see Introduction). So far, evidence for the functional need of inter-repeat cohesion came from the findings that (i) most purified yeast and vertebrate FG repeats can form three-dimensional meshworks (i.e. macroscopical hydrogels) faithfully reproducing the permeability properties of NPCs *in vitro* (Frey and Görlich, 2007; Frey and Görlich, 2009; Milles and Lemke, 2011) and (ii) only such cohesive FG repeats can maintain a functional barrier in a physiological context (i.e. NPCs reconstituted in a *Xenopus* egg extract; Hülsmann et al., 2012). But if a selective phase was the essential feature of the NPC permeability barrier, then the cohesiveness of FG repeats should be evolutionary conserved in all Eukaryotes.

In this study, it was now systematically shown that (i) diverse species from all eukaryotic claces indeed contain cohesive FG repeats capable of forming hydrogel-based permeability barriers redolent of NPCs and that (ii) FG repeats incapable of forming hydrogels on their own nevertheless are highly optimized and mutation-sensitive sequences that probably must remain adhesive in order to contribute to the overall hydrogel-barrier performance. These findings hence strongly suggest that (i) cohesiveness, like NTR binding, is a highly conserved and essential feature of barrier-forming FG repeats and that (ii) this strong evolutionary pressure reflects the necessity of every NPC permeability barrier to rely on cohesive FG repeats to form a hydrogel-like selective phase.

Notably, these conclusions are well affirmed by several other recent findings. First, the major emphasis of the present work was on FG repeats derived from Nup98 homologs, the nucleoporin that was lately identified as the essential component of the (vertebrate) NPC permeability barrier (Laurell et al., 2011; Hülsmann et al., 2012). Second, also in the yeast *Saccharomyces cerevisiae*, the three Nup98 homologs play a dominant role, as already the genomic deletion of the Nup100 and Nup116 FG repeats causes lethality, which notably cannot be rescued with any other yeast FG repeat (M. Ridders, personal communication). Third and nevertheless, not only yeast and vertebrates contain multiple other cohesive FG repeats apart from their Nup98 homologs (Frey and Görlich, 2007; 2009 and A. Labohka, personal communication), but also *Trypanosoma* (preliminary data), commonly regarded as one of the most earliest-branching eukaryotes (see Figure 2.1.C and Cavalier-Smith, 2010a).

3.1 Evidence for a continuum of cohesive FG repeats between two extreme modes of inter-repeat interaction with different side effects.

However, if the primary task of FG repeats inside the nuclear pore is indeed to form a sieve-like hydrogel-based permeability barrier, why then do not all Nup98-derived FG repeats form hydrogels that are perfect barriers *in vitro* (see Figures 2.6 and 2.7)? A plausible answer to this question is that the cohesiveness of individual FG repeats evidently comes with intrinsic 'side effects'. Specifically, the eukaryote-wide sequence analysis of FG repeats and the data on the permeability properties of FG hydrogels presented in this study suggest that there are two extreme solutions for rendering FG repeats cohesive, which cause different problems.

On the one hand, solution (1) seems to be possessing great hydrophobicity, as e.g. the Nup98-derived FG repeats of chordates or plants (see Figure 2.3). Indeed, these repeats formed tight hydrogels that were excellent passive barriers (Figure 2.6), but however hardly sustained the influx of NTR•cargo complexes (Figure 2.7). Solution (2) on the other hand presumably is being extremely NQ-rich and prone to engage in amyloid-like interchain β-sheets, as e.g. a number of yeast FG repeats or the mircronuclear *Tetrahymena* IF repeats (Figure 2.3). Yet, hydrogels constituted from these repeats showed an 'incorrect' selectivity, as they also had an undesired affinity for inert molecules such as MBP-mCherry (Figure 2.7).

Interestingly, the different cohesiveness modes of the FG repeats also translated into opposite technical problems with the handling of the proteins during hydrogel formation. Two questions are important in this regard. First, how fast does a repeat gelate? And second, how compact will the final gel be?

The hydrophobic FG repeats from human or *Arabidopsis* for example are so cohesive that they require a low pH and considerable concentrations of GHCl (up to 2M) to prevent an immediate collapse of the lyophilized protein upon solvent addition. In addition, they are at the verge of undergoing a complete phase-separation into a protein-rich phase with strong barrier effects and an aqueous solvent phase.

In contrast, the very NQ- rich *Tetrahymena* IF or *Saccharomyces* $Nsp1_{2-175}$ FG and Nup145 GLFG repeats for instance are nicely soluble even in physiological buffers, allowing the formation of homogenous, transparent hydrogels.

Taken together, both extreme modes of cohesiveness thus give rise to opposite challenges: solution (1) poses the problem that the tightness of the barrier has to be regulated, whereas extreme (2) requires passivation of the barrier. Note that the latter problem was also faced when attempting to convert the NQ-rich Sup35 PrD into a functional FG repeat (see Section 2.3), further illustrating that unspecific background binding of inert molecules is favored by amyloid-like higher-order assemblies of non-globular, NQ-rich and moderately hydrophobic proteins. Additional direct evidence for the notion that there is more than one possible way of making FG repeats cohesive comes from the structural characterization of FG hydrogels by the means of ssNMR: contrary to the previously reported Nsp1 hydrogels (Ader et al., 2010), the *Xenopus* Nup98 hydrogels hardly contain any amyloid-like β-strands (A. Labohka, D. Görlich and M. Baldus, unpublished results) and hence must be cohesive due to other means.

3.2 Remedies for the side effects of FG repeat cohesiveness in the NPC.

Of course, the notion that FG repeats make use of different modes of cohesiveness with distinct side effects raises the question which strategies are used to optimize the permeability properties of NPCs *in situ*. In case of the vertebrate pore, a plausible explanation has already been suggested (A. Labohka, unpublished results). Here, it is well know that FG repeats such as Nup98 are post-translationally modified with O-linked sugar chains (see Section 1.2).

Remarkably, O-GlcNAc-modified *Xenopus* Nup98 FG repeats form transparent hydrogels that fully reproduce the permeability properties of NPCs, i.e. pose firm barriers towards the influx of inert molecules, but allow the efficient uptake and intra-gel diffusion of NTR•cargo complexes (Hülsmann et al., 2012). Whereas the former was also true for unmodified Nup98, the latter was explicitly not observed in the absence of the O-GlcNAc modification (see Figure 2.7). For comparison, the hydrogels formed by the O-GlcNAc-modified Nup98 FG repeats showed a similar selectivity as the hydrogels formed from *Caenorhabditis* Nup98-derived FG repeats in this study (compare Figures 5D and E from Hülsmann et al., 2012 and Figure 2.7). In conclusion, the post-translational addition of highly water-soluble sugar chains is a well suited means of weakening the inter-repeat interactions between highly hydrophobic FG repeats to prevent the formation of a meshwork that is too tight for NTRs to efficiently pass (A. Labohka, unpublished).

In agreement with this notion, the enzyme required for this kind of modification, O-GlcNAc transferase (reviewed in Hurtado-Guerrero et al., 2008), is also found in plants such as *Arabidopsis*, rice or wheat (Table 3.1), which together with vertebrates contain the most hydrophobic FG repeats. The enzyme is moreover found in worms, flies and fish, as well as some fungi, Amoebozoa and Chromaleveolates (Table 3.1). However, it remains to be tested if Nups are indeed glycosylated in these species and if so, how this influences the permeability properties of their NPCs. Of particular interest in this regard is the worm *Caenorhabditis elegans*, which apparently already contains optimally balanced FG repeats without the addition of sugar moieties.

Table 3.1. *Homologs of OGT identified in Eukarya.*

Clade	Supergroup	Homolog identified?
	Euglenozoa	X
Excavata	Parabasalids	X
	Heterolobosea	X
	Diplomonads	**Yes**
Rhizaria	Radiolaria	X
	Cercozoa	X
	Foraminifera	X
Chromalveolata	**Alveolates**	**Yes**
	Stramenophiles	**Yes**
	Haptophytes	X
	Cryptophytes	X
Plantae	**Plants**	**Yes**
	Chlorophytes	**Yes**
	Rhodophytes	X
	Glaucophytes	X
Amoebozoa	**Slime Moulds**	**Yes**
	Entamoebae	X
Opisthokonta	**Vertebrates**	**Yes**
	Diptera	**Yes**
	Tunicates	**Yes**
	Nematodes	**Yes**
	Choanoflagellates	**Yes**
	Fungi	**Yes**

Yeast FG repeats in contrast are knowingly not O-GlcNAc-modifed and rely on a different mode of cohesiveness, which is characteristic of amyloid-like interchain β-strand formation involving their NQ-rich spacers. Primarily, the problem of NQ-rich FG repeats is that they also bind inert molecules. Additionally, they face the danger of forming amyloid-like aggregates as the Sup35 PrD (Section 2.3) rather than hydrogels (Halfmann et al., 2012b).

As shown in Section 2.2, the background binding of macromolecules such as MBP-mCherry to hydrogels formed by NQ-rich FG repeats can however be repressed when they are blended with highly charged, non-cohesive FG repeats. This situation is reminiscent of the need for 'passivation' commonly encountered by biochemists, e.g. when coupling proteins such as antibodies to surfaces for the preparation of affinity matrices. Remarkably though, also the entry and intra-gel diffusion of active species is facilitated in response to the passivation by non-cohesive FG repeats. Please not that such a fine-tuning of barrier properties via blending is also well known from similar cellular permeability barriers like lipid bilayers, where the ratio of different lipids to each other analogously determines the permeability for given molecules (Voet and Voet, 2010).

Two experiments performed in the present study indeed further suggest that FG repeat blending, like O-GlcNAc modification in higher eukaryotes, might be a general mechanism for the optimization of hydrogel-based permeability barriers. First, also the asparagine-rich micronuclear *Tetrahymena* Nup98 repeats form hydrogels with background binding problems that can be alleviated by including e.g. the highly charged FSFG repeats from Nsp1 (see Section 2.2.2). Second, fusing the charged Nsp1 repeats to the NQ-rich Sup35 PrD mutants, which can be regarded as conceptualizations of NQ-rich FG repeats, allowed to create hybrid diblock polymers that formed hydrogels behaving identical to WT Nsp1 gels. On their own, the Sup35 PrD mutants however only formed imperfect barriers (see Section 2.3.5).

Importantly in regard of the blending model, the passivating FG repeats (as e.g. the $Nsp1_{274-601}$ domain) seem to be highly optimized sequences that apparently must be well balanced and capable of interacting with the barrier-forming FG repeats (e.g. the $Nsp1_{2-175}$ domain) in order to improve the selectivity of FG hydrogels (see Section 2.2). Otherwise, neither is background binding suppressed, nor the barrier passage of active species facilitated. This notably further highlights the importance of inter-repeat interactions in the context of the NPC permeability barrier, even though individual FG repeats are not cohesive enough to form hydrogels on their own.

Taken together, how can the barrier-enhancing effect of the adhesive FG repeats be explained? A recent biophysical characterization of the full-length Nsp1 hydrogel suggested that despite its macroscopic homogeneity, it represents a composite material that combines a flexible and unstructured microphase formed by the highly charged $Nsp1_{176-601}$ domain and a rather rigid scaffold contributed by the $Nsp1_{2-175}$ portion (Petri et al., 2012).

Macromolecules are consequently believed to preferentially pass through the flexible microphase, which is (i) less protein-rich than the rigid parts of the composite and hence provides a larger volume freely available for unhindered diffusion (of both small inert and larger active species) and (ii) can supposedly be much easier rearranged by NTRs to allow the facilitated translocation of even large cargoes. The $Nsp1_{176-601}$ domain however is not only thought to enhance the permeability of the composite gel, but also to prevent the collapse of the entire FG domain into tightly packed aggregates.

In combination with the previously discussed findings, this suggests three types of inter-repeat interactions in such composite hydrogels: (1) hard to dissolve amyloid-like interchain β-strand interactions that render the selective phase kinetically stable, (2) heterologous interactions between FG motifs in the gel scaffold and FG motifs of the passivating repeats which prevent the collapse of the gel and require NTRs to be transiently opened, and (3) dynamic interactions between the passivating repeats, which likely re-organize without the help of NTRs, but can be stabilized by NTRs to further tighten the barrier (as observed in Frey and Görlich, 2009).

One would hence predict that the meshes formed by the latter interactions are easy to open for individual NTRs, whereas the former interactions are much tighter and presumably require the concerted action of multiple NTRs to be dissolved. Also, the disruptive effect of aliphatic alcohols such as (cyclo-)hexanediol (Ribbeck and Görlich, 2002) can likely not be explained by the disruption of the tight amyloid-like interactions within the scaffold microphase (once they are formed), but rather by collapsing the flexible microphase with its presumably mainly hydrophobicity-driven interactions. This notion can be directly tested in future work e.g. by studying the effect of hexanediol on hydrogel-based permeability barriers formed by FG repeats of different cohesiveness types, hence possibly allowing to biochemically prove the different ways of forming selective phases and thus complementing the structural analysis of hydrogels.

Interestingly, when the adhesiveness of the flexible microstructure is (artificially) reduced by the introduction of mutations (see Section 2.2.3), the sieving effect of the hydrogel is abolished, most likely due to the lack of interactions (2) and (3). Yet, in such a scenario, one would expect to find a similar composite hydrogel with the available structural methods, once more highlighting the major future challenges of the field: the need for structural insight, but the lack of suitable methods with near-atomic resolution.

In summary, depending on the exact stochiometry of cohesive vs. adhesive FG repeats, the permeability of the barrier can hence be fine tuned by providing more or less easy to open meshes.

3.3 Can FG repeats with different types of cohesiveness be functionally exchanged in the context of the NPC?

An interesting question arising from the idea that there are different solutions for making FG repeats cohesive is whether the extremes of the continuum are interchangeable in the context of the NPC without disrupting the passive exclusion or active transport of selected molecules. Two recent studies allow to speculate on this. On the one hand, the permeability barrier of NPCs reconstituted *in vitro* from *Xenopus* egg extracts depleted of endogenous FG repeats can indeed either be rescued by the addition of highly hydrophobic (e.g. *Xenopus*) or NQ-rich (e.g. *Saccharomyces*) Nup98-derived FG repeats (Hülsmann et al., 2012). On the other hand however, a *Saccharomyces cerevisae* strain genetically devoid of all three endogenous Nup98 homologs (i.e. ScNup100, ScNup116 and ScNup145N) is not viable, but can be very well rescued e.g. by plasmid-based copies of the NQ-rich FG repeat-containing ScNup100 or ScNup116, whereas the *Xenopus* or human Nup98-derived FG repeats (fused to the ScNup100 or ScNup116 anchor domains) fail to do so (M. Ridders, unpublished results). What does this apparent ambiguity tell us?

First, the successful *in vitro* reconstitution of NPCs excluding inert molecules and mediating active nuclear import with both kind of repeats conceptually proves that both extreme solutions support functional permeability barriers, even when the NPC scaffold varies in size and its basic building blocks.

Second, in the case of the reconstituted NPCs, only active import of a fluorescently-labelled IBB-MBP cargo was tested as a basic proof of principle. In the much more stringent *in vivo* system however, all nuclear import and export pathways have to be supported, including export of gigantic cargoes such as ribosomal subunits or RNP particles. Clearly, this involves much more NTRs and requires the opening of much more individual meshes at once, which at such a large scale might become structurally and/or energetically problematic when the barrier principle is so severely altered. In other words, the cohesiveness of the different repeat types might not be equivalent.

Third and presumably tightly linked with the previous point, the respective mechanisms used to solve the side effects associated with either type of FG repeat cohesiveness might not be fully compatible (see also discussion above). In this regard, the localization of the different FG repeats in the NPC relative to each other likely is of crucial importance and adapted to the respective endogenous FG repeat composition.

Fourth, individual components of the nuclear transport machineries might be optimized for either type of repeat (as e.g. certain *Tetrahymena* NTRs or import adaptors are, see below).

Fifth and taken together, despite the significant advances in understanding the basic principles of the NPC permeability barrier and its building blocks, our knowledge of the selectivity mechanism still lacks molecular and ultimately thorough structural detail.

3.4 The relationship between NQ-rich FG repeats and yeast prions.

Most yeast prions are NQ-rich proteins prone to form amyloids, the latter which are ordered aggregates driven by the formation of interchain β-sheets. As discussed in Section 2.2, such prions resemble NQ-rich FG repeats, as they both share a similar amino acid sequence bias and show the same structural features upon congregation into either amyloids or hydrogels.

Interestingly in this regard, the data presented in this study (Figure 4.3) suggest that the nature of the predominant hydrophobic amino acid seems to primarily govern whether NQ-rich sequences aggregate into highly ordered, water-free amyloid fibrils (when tyrosine is favored) or water-retaining hydrogels (when phenylalanine is preferred). In agreement with this finding, a mutant version of full-length Nsp1, where all endogenous phenylalanines have been exchanged to tyrosine residues, does no longer form transparent, but rather opaque hydrogels (Frey et al., 2006), which is indicative of macroscopic inhomogenities. Presumably due to the high density of charges in the (also F-Y converted) $Nsp1_{274-601}$ domain however, the mutant does not completely collapse into amyloid-like aggregates.

In fact, it is well known in the prion field that charges, either in particular positions of an amyloidogenic stretch or in an excess surrounding a sticky site, serve as 'amyloid breakers' (López de la Paz and Serrano, 2004). Moreover, also the presence of proline residues, which highly disfavors the formation of β-strands, is considered as amyloid breaking (Adessi and Soto, 2002). Hence it is not surprising that each of the 17 regular repeat units in the $Nsp1_{274-601}$ domain also contains at least one proline.

Amyloid aggregates are commonly associated with pathological conditions (Chiti and Dobson, 2006). Thus, great efforts are made to devise therapeutic strategies that allow interfering with amyloid formation. Notably, some promising approaches focus on charged, proline-rich peptides to destabilize amyloids *in vivo* (Adessi and Soto, 2002; Permanne et al., 2002), hence unwittingly trying to mimick the situation in the nuclear pore (of lower eukaryotes), where equivalent FG domains similarly enhance the overall barrier performance (see above). Taken together, the proceedings of prion research further support the FG repeat blending model introduced here.

In the context of FG repeats however, amyloid-breaking residues such as proline or charged amino acids might not only prevent FG repeat aggregation, but also stimulate NTR binding, e.g. as a result of the (local) absence of secondary structure around an FG motif or an enhanced affinity for the FG motifs in that domain (perhaps due to favorable electrostatic interactions, as suggested in (Colwell et al., 2010).

Interestingly, a series of recent studies showed that, apart from the bias in the hydrophobic side chains and the presence of amyloid breakers, the amyloidogenic potential of a given disordered protein is also decided on by the ratio of asparagine to glutamine (Alberti et al., 2009; Halfmann et al., 2011). This finding was unexpected, as previous studies assumed that both residues, which only differ by one methyl group equivalent in their total side chain length, contribute equally to prion formation (e.g. (Michelitsch and Weissman, 2000). However, N-rich prions clearly show a higher tendency to form amyloids, whereas Q-rich proteins preferentially form highly cytotoxic non-amyloid aggregates (Halfmann et al., 2011). Supposedly, Q-rich prions convert far less efficiently from a soluble oligomeric intermediate form into ordered amyloids. Indeed, polyQ proteins take much longer to aggregate and oftentimes require a prior extension of their polyQ stretches (M. Aksu, personal communication and Osherovich et al., 2004). The explanation for the inequality of the two amide residues given by (Halfmann et al., 2011) can be interpreted such that glutamine is less cohesive than asparagine, as the shorter side chains of the latter favor additional hydrogen bonding to the polypeptide backbone. Please note that in most FG repeats studied here, the more sticky asparagine dominates over glutamine. In the Sup35 PrD, arguably the best studied prion, the opposite is true. Given the findings that (i) Impβ was able to diffuse much better in the hydrogels formed from the Sup35 PrD mutants and (ii) the background binding of inert molecules was slightly reduced as well (see Figure 4.4), an optimal balance of asparagine and glutamine residues might matter for the design principles of the NPC permeability barrier as well. In particular, these findings shine a new light on the observation that the FG repeats of the Nup98 homologs 'sealing' the micronuclei of ciliates are almost exclusively made up of asparagine residues (see below).

The other way around however, it also valid to ask how the nuclear transport field can fuel prion and amyloidosis research, especially in regard to novel therapeutic approaches. As mentioned above, the primary problem here is to dissolve toxic prion aggregates, which can either be non-amyloid confomers or *bona fide* amyloid fibrils according to the current discussions in the field. Intriguingly, NTRs are capable (or at least highly predisposed) of doing that in the nuclear pore. Additionally given the structural knowledge of NTR-FG repeat interactions and amyloid formation, as well as the availability of powerful new optimization techniques such as *in vitro* evolution, it hence may be well possible to specifically design amyloid-dissolving agents based on NTR scaffolds. Prime starting candidates would be the *Tetrahymena* NTRs with known preference for the asparagine-rich micronulcear Nup98 repeats (see below). Alternatively, also the MIC-specific *Tetrahymena* import adaptors should be considered, which might be even less prone to interfere with endogenous nucleocytoplasmic transport once optimized for amyloid breaking.

3.5 Is the Sup35 Y-FG PrD a functional FG repeat?

Given the relationship between yeast prions and FG repeats outlined above, a goal of this study was to convert the Sup35 PrD into a functional FG repeat. Two major achievements have been made in this regard. First, as a result of changing all tyrosines of the Sup35 PrD into phenylalanines, its propensity to assemble into fibrillar aggregates was abolished. The Y-F mutant PrD rather formed hydrogels (Figure 4.3). Second, by introducing FG motifs, a broad affinity for NTRs was established (Figure 4.4). Yet, can the Sup35 Y-F and Y-FG PrDs be considered functional FG repeats?

To ultimately answer this question, it remains to be seen how they behave in the context of the NPC, in particular whether they fully replace endogenous FG repeats in maintaining the NPC permeability barrier. Within the yeast NPC, a number of FG Nups contain NQ-rich FG domains, including primarily Nsp1, Nup49, Nup57, Nup100 and Nup116. Hence, these sites are prime locations to test the effect of the mutant Sup35 PrDs on the yeast NPC permeability barrier.

A very stringent *in vivo* assay to address such questions was recently developed in our lab (M. Ridders, personal communication). It relies on a yeast genetic background where enough FG repeat mass has been genomically deleted to make a plasmid-based copy of a given FG Nup essential for proper NPC function and viability. A synthetic construct containing any FG repeat region fused to the anchoring domain of the complementing FG Nup can then additionally be transformed into the yeast. When now counter-selecting against the wild-type FG Nup-bearing plasmid with the drug 5-Fluorootic acid (abbreviated FOA), the integrity of the NPC permeability barrier and viability of the yeast will entirely depend on whether the synthetic construct can also rescue the otherwise lethal deletions.

Given the comparably short size of the Sup35 PrD and the Nup57 FG repeats, we first asked if the modified prions can restore NPC function and viability in a genetic background where Nup57 and the FG repeats of Nsp1 and Nup49 were deleted. Preliminary results indicate that this is not the case (M. Ridders, personal communication). However, recent findings suggest that the preferred locations for NQ-rich FG repeats within the yeast NPC are the Nup100 and Nup116 anchor sites (M. Ridders, personal communication). Note that the FG repeats from Nup100 or Nup116 indeed also cannot rescue lethality when fused to the Nup57 anchor point.

Additionally, it will be interesting to see whether the Sup35 Y-F and Y-FG PrDs can also functionally replace the $Nsp1_{2-175}$ domain *in vivo*, as suggested by the hydrogel experiments presented above. Thus, by systematically testing different genetic backgrounds and anchoring locations, future work will not only clarify if the Sup35 Y-FG PrD indeed resembles *bona fide* NQ-rich FG repeats, but also help to understand which features of FG repeats in general are essential for NPC permeability barrier function *in vivo*.

3.6 Sequence determinants of FG repeats.

In various parts of the present study, valuable information on the sequence determinants of FG repeats and their functional implications on the permeability properties of FG repeat hydrogels were collected. Taken together, FG repeats consists of mainly glycine, alanine, serine, threonine, proline and phenylalanine residues. Generally, threonine is more dominant than serine, presumably because it is more hydrophobic. To make FG repeats cohesive, either more hydrophobic amino acids, preferentially leucine (but no aromatics other than phenylalanine), or asparagines and glutamines are included. Especially the asparagine side chains promote the formation of amyloid-like interchain β-strands between opposing FG repeats. Charged residues (i.e. almost exclusively lysine) are primarily required to titrate out the cohesiveness of FG repeats, but may also favor NTR binding. Ideally, an optimal solution between overall hydrophobicity and NQ-content is found, as is the case for the *Caenorhabditis* repeats.

However, to fashion an FG repeat *de novo*, it certainly will not be enough to randomly choose from these preselected amino acids. How often would one have to shuffle a sequence of with an amino acid bias typical of FG repeats to really obtain a functional, barrier-forming protein? How long does an NQ-rich stretch have to be in order to promote cohesiveness? What degree of hydrophobicity is necessary? And how extreme can an FG repeat locally be, given it is globally balanced? In summary, we are only beginning to understand the exact design principles underlying cohesive FG repeats. This is perhaps best illustrated by the finding that the extremely charge-rich $Nsp1_{274-601}$ FSFG repeats can dilute the cohesiveness of the macronuclear *Tetrahymena* Nup98 GLFG repeats such that they do not form hydrogels anymore, but are not sufficient to fully suppress the background binding of MBP-mCherry to the micronuclear IF repeats (see Section 2.2.2).

To answer these questions and test basic ideas, future work could focus on interconverting FG repeats from one mode of cohesiveness to the other. An interesting starting point for such investigations would be the non-cohesive $Nsp1_{274-601}$ domain: which sequence features are required to make it cohesive in the one or the other way? Preliminary results suggest that reducing the charge density without increasing the overall hydrophobicity is not sufficient to make it gelate on its own. Other barrier-forming model FG repeats rich in charged amino acids, such as the $Nup153_{877-1122}$ domain, should be very helpful in this respect. Note that, similar to the correlation made for the Nup98-derived FG repeats, also here the number of FG repeat-untypical hydrophobic amino acids is elevated to increase the overall hydrophobicity, e.g. due to additional valine or leucine residues in front of the FG motifs.

Eventually, these efforts will hopefully contribute to a more detailed understanding of the sequence effects on the structures functions adopted by non-globular proteins in general and FG repeats in particular. With that knowledge, bioinformatic tools for the *in silico* prediction of FG repeats and improved computational models of their behavior can finally be compiled (K. Kirli, personal communication).

3.7 On the evolution of the NPC and its permeability barrier.

NPCs are the largest marcomolecular complexes of eukaryotic cells, build from multiple copies of ~30 different Nups. How did such an elaborate structure emerge in evolution? In the past, numerous comparative genomic studies have suggested that all major Nup subcomplexes are traceable to the last common eukaryotic ancestor (LCEA) (Mans et al., 2004; Bapteste et al., 2005; Neumann et al., 2010). Generally however, only the structures of the Nups are highly conserved, whereas there is much more divergence on a sequence level. Especially in the case of the non-globular FG repeats, which *en masse* do not fold in a defined manner, the sequence divergence is enormous, although their amino acid composition is mostly conserved (Denning and Rexach, 2007 and this study). Two different 'tempos of evolution' were hence proposed for the NPC, allowing for highly adapting building blocks in an ancient fixed structure (Bapteste et al., 2005).

As stated in the beginning of the Discussion, the present study contributed in two ways to our understanding of the NPC evolution. First, the finding that FG repeats from all eukaryotic clades form hydrogels reproducing the fundamental aspects of passive and facilitated nuclear transport is compelling evidence for the notion that the selectivity mechanism of the NCP permeability barrier is conserved from the LCEA. Second, this strongly suggests that the selective pressure forcing FG repeats to maintain their characteristic amino acid composition is not only the ability to bind NTRs, but at least in equal contribution also the need to remain cohesive. The continuum of FG repeats, ranging from highly hydrophobic to NQ-rich, likely marks the evolutionary tolerated regime in this regard.

Putting these results into context and summarizing the literature, the following sequence of basic, independent co-evolutionary events for the emergence of NPCs seems plausible: (i) obtaining a closed nuclear envelope shielding the DNA with proto-NPCs perforating the new endomembrane and (ii) equipping the proto-NPC with an FG repeat-based permeability barrier that can be traversed by Ran-regulated proto-NTRs. The underlying rationale in both cases is based on the simple fact that the genetic information cannot be sequestered away in a novel compartment without ensuring constant access to it, either direct or regulated.

The origins of the nuclear membrane are highly disputed. Membrane heredity, one of the most fundamental principles of biology, asserts that biological membranes cannot be created *de novo*, but must be derived from pre-existing membranes (see e.g. Cavalier-Smith, 2001). Hence, two questions arise: (i) where did the nuclear membrane originate from and (ii) what was the preceding event or selective pressure driving this acquisition? Possible answers to the former question are an endosymbiotic source or an autogenous origin from an own membrane. However, it is highly unlikely that a protoeukaryote took up a bacterium, archaea or inferior protoeukaryote, 'enslaved' the foreign cell and eventually shuffled its own genetic information in the membrane shell thus obtained, let alone that the endosymbiont took over controlling the host. Indeed, it is much more convincing that the nuclear envelope originated from invaginations of the protoeukaryotic plasma membrane. The recent identification of a bacterial superphylum containing intracellular membranes strongly supports this notion (reviewed in Fuerst, 2005). Recall that bacteria rely on plasma membrane-associated motors to segregate duplicated DNA. If by chance a patch of the membrane to which DNA is linked was now accidentally invaginated, then the first step towards forming a nucleus would have been already taken (Cavalier-Smith, 2010b).

Whatever the selective pressure for acquiring the nuclear membrane however really was, it remains without doubt that the NPC scaffold with its membrane-capping and pore-forming ability must have already been invented before the complete enwrapping of the genetic information was possible. The protocoatomer hypothesis suggests a common evolutionary origin for the NPC scaffold and clathrin, COPI and COPII vesicle coats (Devos et al., 2004; Degrasse et al., 2009; Field et al., 2011). Remarkably, such 'membrane-coat' proteins have also been identified in the bacteria containing intracellular membranes (Santarella-Mellwig et al., 2010), suggesting that the fundamental requirements for the evolution of the nucleus were indeed already present in a possible eukaryotic cenancestor of bacterial origin.

Also the second concerted step of the NPC evolution, the actual acquisition of the permeability barrier, was a far-reaching event of eukaryogenesis. In fact, it not only set the basis for regulated exchange between the cytoplasm and the newly emerging nuclear compartment, but also was a necessary prerequisite for changing the genome organization from an operon- into an intron/exon-based arrangement and allow alternative splicing to increase the genomic coding potential (Cavalier-Smith, 2010b).

However, in contrast to the scaffold Nups, particularly the evolutionary origins of FG repeats remain elusive. Logically, it seems plausible that the protoeukaryote relied on multiple copies of only few, if not just one, cohesive precursor FG repeat. Most likely, these precursors featured (the eventually name-giving) phenylalanines because of three reasons. First, both the hydrophobic and aromatic stacking interactions between phenylalanines of opposing FG motifs are presumably energetically required to allow for the formation of kinetically stable meshes (unlike non-aromatic, hydrophobic side chains), which nevertheless can be easily opened by NTRs. Second, for similar steric/energetic reasons, aromatic side chains might be easier to recognize for NTRs in comparison to other hydrophobic residues (see also the crystal structures of the Impβ-FG motif interaction in this regard, Bayliss et al., 2000; 2002b). Third, as shown in Section 2.3, phenylalanines, other than e.g. tyrosines, are less likely to promote the unfavorable aggregation of FG repeats.

By the time of the LCEA however, the NPC presumably already contained primordial versions of Nup98 and the Nup62 subcomplex (Neumann et al., 2010). Next in line, the earliest-branching eukaryotes (i.e. the excavates) embrace both hydrophobic and amyloid-like interactions as the basis for their cohesiveness, as judged by the cluster analysis of the Nup98 database performed in the present study. Considering that, based on their position in the eukaryotic tree, excavates have had the most time to explore the allowed boundaries for FG repeat cohesiveness, this finding makes sense, but also highlights how hard it is to predict which extreme solution to confer cohesiveness had first been employed by the proto-NPC. Note that the same logic, in combination with their crucial role in the (vertebrate) NPC permeability barrier, also explains why the FG repeat regions of the vertebrate Nup98 homologs are so strikingly similar.

3.8 The current understanding of nuclear dualism.

Looking at the evolution of cellular biodiversity, two major trends can be detected: thriving towards more, or seeking less complexity. A remarkable example of the former strategy are Ciliates, which comprise a group of highly sophisticated unicellular eukaryotes that contain both a 'somatic' macro- and a 'germ-line' micronucleus in a common cytoplasm, a phenomenon referred to as nuclear dualism. To maintain their specific functions, both nuclei require different subsets of proteins, which in turn have to be selectively imported by adapted nuclear transport machineries (see Introduction). In general, the nuclear transport machinery comprises NPCs with FG repeat-based permeability barriers, NTRs and the RanGTP-system. Nuclear dualism however demands for two different permeability barriers with orthogonal specificity, in combination with compatible NTRs.

Three arguments propose that the *Tetrahymena* Nup98 homologs are the key barrier-forming FG repeats of the two antipodal ciliate NPC permeability barriers. First, they are the only known nucleus-specific permeability barrier components of the model ciliate *Tetrahymena* (see Table 1.3 and references therein). Second, both nuclei each contain either two macro- or micronuclear Nup98 orthologs (Malone et al., 2008; Iwamoto et al., 2009), indicating their relative abundance in comparison to other Nups. Third, a picture is emerging in which Nup98 is the essential component of the NPC permeability barrier, at least in vertebrates (Hülsmann et al., 2012) and yeast (M. Ridders, personal communication). Yet, how can the different *Tetrahymena* Nup98 homologs compose orthogonal barriers?

In light of the preceding discussion of the evolutionary conserved principles that allow FG repeats to form hydrogel-based permeability barriers with NPC-like properties, a plausible model would be to have two types of sieve-like meshworks governed by (structurally and energetically) distinguished meshes that can only be opened by specialized NTRs. In agreement with this view, the results presented in this study suggest that the *Tetrahymena* Nup98 homologs fundamentally differ in their cohesiveness modes, which are either based on hydrophobicity (MAC) or amyloid-like β-sheet formation (MIC). In fact, the micronuclear Nup98 repeats are an extreme example for non-globular, NQ-rich proteins (see Figure 2.2). As noted before (Iwamoto et al., 2009), the MAC and MIC repeats additionally also differ in the flavor of their hydrophobic motifs: whereas the MAC repeats contain canonical GLFG motifs, the MIC repeats feature atypical IF motifs.

Recall in this regard that an *in vivo* exchange of the barrier-forming FG repeats with their respective counterparts of the opposite cohesiveness type, even if they do not vary in their FG motif flavor, was indeed not tolerated in yeast (personal communication with M. Ridders, see above). Analogously in *Tetrahymena*, MAC species could no longer be imported when the MAC was equipped with MIC repeats and vice versa, MIC species could not access a MIC equipped with MAC repeats (Iwamoto et al., 2009).

Taken together, this strongly suggests that Ciliates first and foremost make use of the two different fundamental principles to build hydrogel-based permeability barriers in order to operate distinct NPCs with orthogonal specificity in a common cytoplasm.

Notably, the two nuclei vary considerably in their size. In *Tetrahymena*, the somatic MAC has a volume of ~500µm, whereas the germ-line MIC only takes up ~10µm (Iwamoto et al., 2009). However, both the MAC and MIC nuclear membrane contain ~45

NPCs/µm. Hence, the MIC contains ~3-times more 'entry gates' per µm of nuclear volume, *per se* resulting in a greater nuclear import capacity compared to the MAC. This consideration suggests that access to the MIC via the micronuclear NPCs must primarily be restricted for most NTR•cargo complexes, whereas the macronuclear NPCs are rather busy with keeping up the extensive bidirectional exchange between the MAC and cytoplasm.

In light of the finding that in particular N-rich prions readily form amyloids (see discussion above), i.e. tightly zip-up along the axis of their polypeptide backbones, it is hence tempting to speculate that the presumably highly abundant and extremely N-rich micronuclear IF repeats indeed 'seal' the micronuclear NPCs and that only specialized NTRs can open this zipped curtain again. In case of *Tetrahymena*, a promising candidate for the latter is TtImpB6: it enriches more than 3-fold better in MIC compared to MAC hydrogels and is far better taken up by MIC hydrogels than the MAC-specific NTR TtImpB3 (see Figure 5.3). Note however that the MIC repeats, like other NQ-rich FG repeats such as yeast $Nsp1_{2-165}$ or Nup145, form hydrogels that show a significant background binding of inert molecules (see Figures 2.7 and 3.3), thus questioning the idea that the micronuclear NPCs are shuttered by the IF repeats.

Similarly, also reconstituting the MAC permeability barrier with MAC hydrogels was not yet successful, as the gels hardly allowed for the NTR fluxes expected for the high-performance barriers required by the transcriptionally active MAC (see Figure 2.7).

Hence, although the *Tetrahymena* Nup98 homologs seem to dominate the permeability barriers of the MAC and MIC (Iwamoto et al., 2009), they do not seem to be sufficient to form adequate orthogonal barriers on their own. The finding that FG repeat blending in principle can improve the performance of the MIC hydrogels as well (see Figure 3.3), further supports the view that additional FG repeats likely contribute to the overall properties and specificity of the macro- and micronuclear permeability barriers.

How would one expect these FG repeats to look like? In general, the macronuclear pore supposedly is filled with additional hydrophobic, and the micronuclear pore with further NQ-rich FG repeats. In addition, more unconventional FG motifs like the IF signature of the micronuclear Nup98 repeats are probably to be found in novel MIC repeats. Indeed, the initial attempts to complete the inventory of *Tetrahymena* FG repeats with the help of the high-affinity $Imp\beta_{45-462}$ fragment (see Figure 5.8) revealed that a number of putative FG repeats contained only few *bona fide* FG motifs, but nevertheless formed selective hydrogels (Figure 5.9). However, in future work it remains to be seen whether these candidates indeed localize to either of the two nuclear envelopes and how *Tetrahymena* NTRs respond to them.

Strikingly, amongst the newly identified putative FG repeats featuring canonical FG motifs, the two most prominent hits showed an unusually high net charge, which in the one case was even negative. This finding provokes the speculation whether electrostatic interactions between FG repeats and NTRs, as well as FG repeats and FG repeats, contribute to the selectivity mechanisms in the unique setting of nuclear dualism. Two possible ways are conceivable.

First, repulsive electrostatic interactions between the FG repeats of the barrier and approaching NTR•cargo complexes could prevent mis-targeted species from entering the wrong permeability barrier. This would be an ideal job for non-cohesive FG repeats located at the cytoplasmic verge of the NPC (i.e. the cytoplasmic fibrils), which do not necessarily need to participate in forming the main, sieve-like hydrogel barrier filling the central NPC channel.

Second, one of the two nuclei could make use of attractive electrostatic interactions between neighboring FG repeats to additionally stabilize the central meshwork. Recall that Impβ-like NTRs are superhelical molecules containing two arches that typically 'wrap' around their cargoes. If both arches were oppositely charged, such an NTR would theoretically be suited to also open electrostatically stabilized meshes. Notably, based on a biophysical model, charge has already previously been suggested as a selection criterion for conventional nuclear transport (Colwell et al., 2010), which was however never experimentally verified.

Yet, as this latter theoretical thought once more illustrates, changes to the permeability barrier must always be matched by compatible changes in the NTRs. In agreement with this notion and based on the finding that *Tetrahymena* contains multiple disparate Imp5 homologs, it was suggested here that certain families of cargo-specialized NTRs have expanded and diverged to remain functional in either of the two nucleus-specific transport routes. Moreover, it was shown that at least some of the many Impα-like import adaptors of *Tetrahymena* might not simply be required to broaden the spectrum of recognizable cargoes for a given NTR, but even more importantly also contribute to the overall nuclear specificity of their respective NTR•adaptor•cargo complexes. A prime example is the MIC-specific TtIMA8, which on its own better penetrates into and diffuses within the MIC hydrogels than the above-mentioned TtImpB6.

Another promising candidate that remains to be characterized is a novel putative import adaptor that bears homology to the ARM repeats of Imp-αs and contains terminal EF hand Ca^{2+}-binding motifs. It prominently binds to both MAC and MIC repeats alike and hints at a possible regulation of nuclear transport in *Tetrahymena* by the second messenger calcium, possibly by modulating the affinity of the Impα towards cargo and FG repeats in dependence of its Ca^{2+}-bound state. Possibly, regulated nucleocytoplasmic transport might be involved in coordinating the amitotic division of the MAC with the mitotic division of the MIC prior to cytokinesis. Ciliates contain large pools of Ca^{2+} primarily in the alveolar sacs underneath the plasma membrane and the ER (Plattner and Klauke, 2001). Please also note that Ca^{2+}-regulated nucleocytoplasmic transport has been suggested before, but mainly in connection with (highly debated) Ca^{2+}-induced structural changes in the NPC (Sarma and Yang, 2011). However, all of these hypothesis have to be further tested and ultimately evaluated *in vivo*.

To conclude, what might the evolutionary origin of nuclear dualism be? It is well possible that this unique cell biological phenomenon simply arose to cover up for an error during cell division. Consider that every cell undergoing closed mitosis has to make sure that the two replicated nuclei are equally distributed during cytokinesis. However, at some point during evolution, the ciliate cenancestor could have been left with two nuclei by mistake. A simple possible solution to e.g. prevent gene dosage problems was to shut one of the two nuclei down by plugging its NPCs with NQ-rich IF repeats. Note that the ciliate proteome features many NQ-rich proteins so that presumably plenty of suited candidates where readily available. In connection with the discussion above, this scenario suggests that the different solutions for FG repeat cohesiveness have been independently evolved multiple times. Having now a 'backup' nucleus at hand, the 'innovation rate' of the main nucleus in principle could go up, likely leading to a situation where new innovations further solidified the dual roles of the two nuclei. This new cellular organization might have been particularly advantageous because it allowed ciliates to grow a rather large cell body and in turn to feed from the highly abundant smaller cells in their environment (e.g. bacteria).

4 Materials and Methods
4.1 Materials

4.1.1 Chemicals

If not stated otherwise, all chemical were purchased to the highest available purity from Calbiochem (San Diego, California, USA), Gibco Life Technologies (Paisley, United Kingdom), Merck (Darmstadt, Germany), MoBiTech (Göttingen, Germany), Pharmacia (Uppsala, Sweden), Pierce (part of ThermoScientific, Rockford, Illinois, USA), Promega (Madison, Wisconsin, USA), Quiagen (Hilden, Germanny), Roche Diagnostics (Mannheim, Germany), Carl Roth (Karlsruhe, Germany) or Sigma-Aldrich (St. Louis, Missouri, USA).

4.1.2 Instruments

Centrifuges

5415R tabletop centrifuge	Eppendorf AG, Hamburg, Germany
5424 tabletop centrifuge	Eppendorf AG, Hamburg, Germany
RC6 plus centrifuge	Sorvall (part of ThermoScientific, Waltham, Massachusetts, USA)
WX Ultra centrifuge	Sorvall (part of ThermoScientific, Waltham, Massachusetts, USA)
Discovery M120	Sorvall (part of ThermoScientific, Waltham, Massachusetts, USA)

Rotors

F9 ultracentrifuge rotor	FiberLite series; Piramoon Technologies, Santa Clara, California, USA

F10 ultracentrifuge rotor	FiberLite series; Piramoon Technologies, Santa Clara, California, USA
S45A rotor	Sorvall (part of ThermoScientific, Waltham, Massachusetts, USA)
S55S rotor	Sorvall (part of ThermoScientific, Waltham, Massachusetts, USA)
T647-5	Sorvall (part of ThermoScientific, Waltham, Massachusetts, USA)
T1250	Sorvall (part of ThermoScientific, Waltham, Massachusetts, USA)

Other Instruments

Äkta Purifier	Pharmacia, Upsala, Sweden
Biophotometer-plus	Eppendorf AG, Hamburg, Germany
Climo-Shaker ISF1-X	Kuhner AG, Basel, Switzerland
GenePulser™	BioRad, Hercules, California, USA
iMac 3.1 GHz Intel Core i5	Apple, Cupertino, California, USA
Jesco HPLC system	Jesco, Essex, United Kingdom
Labfors3 Bioreactor	Infors, Bottmingen, Switzerland
Alpha 1-2 LD plus lyophilizer	Christ, Osterode, Germany
NanoDrop ND-2000	Peqlab Biotechnologies GmbH, Erlangen, Germany
Perfection V700 photo scanner	Epson, Tokio, Japan
Pipetman pipettes	Gilson, Middleton, Wisconsin, USA
SensoQuest lab cycler	SensoQuest, Göttingen, Germany
Sonifier 450	Branson, United Kingdom
RVC 2-18 SpeedVac	Christ, Osterode, Germany
TCS SP5 confocal microscope	Leica Microsystems, Mannheim, Germany

4.1.3 Software

Bento 4	FileMaker Inc., Santa Clara, California, USA
EnzymeX	Mekentosj, Aalsmeer, Netherlands
GeneDesigner 2	DNA2.0, Menlo Park, California, USA
ImageJ	National Institutes of Health, USA
Illustrator CS3/5	Adobe Systems, San Jose, California, USA
LASAF	Leica, Mannheim, Germany
Lasergene 8	DNA-Star, Madison, Wisconsin, USA

Mac OS X Vers.10.6.7	Apple, Cupertino, California, USA
Mathematica 7/8	Wolfram Research, Champaign, Illinois, USA
Microsoft Office	Microsoft Corporation, Redmond, Washington, USA
Oligo 6.8	Molecular Biology Insights, Cascade, Colorado, USA
Papers 2	Mekentosj, Aalsmeer, Netherlands
Photoshop CS3/5	Adobe Systems, San Jose, California, USA
Prism 5	GraphPad Software, La Jolla, California, USA
PyMol	Schrödinger, Portland, Oregon, USA
Scaffold 3	Proteome Software, Portland, Oregon, USA
Scrivener 2	Literature and Latte, Cornwall, United Kingdom

4.1.4 Bioinformatics Resources

In the following, only the publicly available databases and bioinformatics resources used in the scope of this work are listed. Detailed sequence analysis was performed with custom-written Mathematica routines, which are described in detail in the Results and Methods sections.

Databases:

Bigelowiella genome database:	genome.jgi.doe.gov/Bigna1
BioNumbers database:	bionumbers.hms.harvard.edu
NCBI protein database:	ncbi.nlm.nih.gov/protein
PDB (Protein Data Bank):	pdb.org
PFAM protein families database:	pfam.sanger.ac.uk
Uniprot database:	uniprot.org
Tetrahymena Genome Database:	ciliate.org

Homology Searches:

BLAST (Basic Local Alignment Search Tool):	blast.ncbi.nlm.nih.gov

Motif Searches:

ScanProsite:	prosite.expasy.org/scanprosite

Secondary Structure Prediction:

COILS:	ch.embnet.org/software/COILS_form.html
Jpred3:	compbio.dundee.ac.uk/www-jpred/

| PSIPRED: | bioinf.cs.ucl.ac.uk/psipred/ |
| SMART: | smart.embl-heidelberg.de/smart/ |

Alignment Calculation and Representation:

| TCoffee | http://tcoffee.vital-it.ch/cgi-bin/Tcoffee/tcoffee_cgi/index.cgi |
| ProtSkin | http://www.mcgnmr.mcgill.ca/ProtSkin/ |

4.1.5 Standard Buffers and Solutions

10x Pfu polymerase buffer:
100mM Tris-HCl pH8.85, 250mM KCl, 50mM $(NH_4)_2SO_4$, 20mM $MgSO_4$

10x PfuS polymerase buffer:
20mM Tris-HCl pH9.0, 250mM KCl, 15mM MgSO4, 100mM $(NH_4)_2SO_4$, 1% Tween20, 1mg/ml BSA

10x Tth polymerase buffer:
100mM Tris-HCl pH8.9, 15mM $MgCl_2$, 1M KCl, 500µg/ml BSA, 0.5% Tween20

10x Ligation buffer:
500mM Tris pH7.5, 100mM $MgCl_2$, 100mM DTT, 10mM ATP, 250µg/ml BSA

Orange-G DNA loading buffer:
10mM Tris-HCl pH8.0, 10mM EDTA pH8.0, 50% Glycerol (w/v), 0.25% Orange-G (w/v)

DNA ladder:
50 ng/µl 1kb-Ladder or 50ng/µl 100kb-Ladder in Orange sample buffer (Gibco, Paisley, United Kingdom).

50x TAE (Tris-Acetate-EDTA) buffer:
242g Tris-base, 57.1ml Acetic acid, 100ml 0.5M EDTA pH8.0, ddH_2O ad 1 liter.

SDS sample buffer:
125mM Tris pH6.8, 3% SDS, 50mM DTT, 1M Sucrose, Bromphenol-blue

10x SDS-PAGE running buffer:
300g Glycine, 60g Tris-base, 24g SDS, ddH$_2$O ad 2 liters.

Coomassie stock solution:
2% (w/v) Coomassie brilliant blue G250 in 50% Ethanol

Colloidal Coomassie stock solution:
0.08% (w/v) Coomassie brilliant blue G250, 1.6% (v/v) ortho-Phosphoric acid, 8% (w/v) Ammonium sulfate, 20% (v/v) Methanol.

1x PBS (pH7.4):
137mM NaCl, 2.7mM KCl, 8.1mM Na$_2$HPO$_4$, 1.76mM KH$_2$PO$_4$

20x E-Mix (energy-regenerating system):
20mM HEPES pH7.5, 250mM Sucrose, 200mM Creatine phosphate, 10mM ATP, 10mM GTP, 1mg/ml Creatine kinase

4.1.6 Cultivation Media

Escherichia coli **media**

2YT liquid medium
16g Tryptone, 10g Yeast extract, 5g NaCl, ddH$_2$O ad 1 liter.

2YT plates
16g Tryptone, 10g Yeast extract, 5g NaCl, 14g Agar, ddH$_2$O ad 1 liter.

LB liquid medium
10g Tryptone, 5g Yeast extract, 10g NaCl, ddH$_2$O ad 1 liter.

LB plates
10g Tryptone, 5g Yeast extract, 10g NaCl, 15g Agar, ddH$_2$O ad 1 liter.

SOB medium
100g Tryptone, 25g Yeast extract, 2.5g NaCl, 0.93g KCl, ddH$_2$0 ad 5 liters.

TB medium
12g Tryptone, 24g Yeast extract, 4ml Glycerol, 2.13g KH$_2$PO$_4$, 12.54g K$_2$HPO$_4$, ddH$_2$O ad 1 liter.

Tetrahymena thermophila media

Modified NEFF medium

0.25% (w/v) Proteose peptone, 0.25% (w/v) Yeast extract, 0.5% (w/v) Glucose, (33.3μM $FeCl_3$).

SPP medium

2% Proteose peptone, 0.1% Yeast extract, 0.2% Glucose, (33.3μM $FeCl_3$).

Starvation medium

10mM Tris-HCl pH7.5

All media were autoclaved prior to use. Before pouring of the plates, 10ml of a 2M K_2HPO_4 stock solution and 10ml of 40% (w/v) glucose were added. For the selection and cultivation of bacteria after transformation or for protein expression, plates and liquid media were supplemented with the appropriate antibiotics in final concentrations of 100μg/ml Ampicillin, 34μg/ml Chloramphenicol or 25μg/ml Kanamycin. To prevent microbial and fungal growth in *Tetrahymena* cultures, 1x antibiotic (effective agents: penicillin, streptomycin)-antimycotic (effective agent: amphotericin B) solution (Gibco, Paisley, United Kingdom) was added.

4.1.7 Strains

Escherichia coli strains

Strains used for cloning:

Strain Name	Genetic Description
TOP10 F'	F' [*lac*I^q Tn10(Tet^R)] *mcr*A Δ(*mrr-hsd*RMS-*mcr*BC) φ80*lac*ZΔM15 Δ*lac*X74 *deo*R *nup*G *rec*A1 *ara*D139 Δ(*ara-leu*)7697 *gal*U *gal*K *rps*L(Str^R) *end*A1 λ⁻
5α F'	F' [*proA⁺B⁺ lac*I^q Δ(*lacZ*)M15 *zzf*::Tn10 (Tet^R)] *fhu*A2Δ(*argF-lacZ*)U169 *pho*A *gln*V44 φ80Δ(*lacZ*)M15 *gyr*A96 *rec*A1 *rel*A1 *end*A1 *thi*-1 *hsd*R17
SURE	*end*A1 *gln*V44 *thi*-1 *gyr*A96 *rel*A1 *lac* *rec*B *rec*J *sbc*C *umu*C::Tn5 *uvr*C e14- Δ(*mcr*CB-*hsd*SMR-*mrr*)171 F'[*pro*AB⁺ *lac*I^q *lac*ZΔM15 Tn10 Amy Cm^R]

Strains used for protein expression:

Strain Name	Genetic Description
BLR	F⁻ *ompT* *hsd*SB(*r*B⁻ *m*B⁻) *gal dcm* Δ(*sr*I-*rec*A)306::Tn10 (Tet^R)
Express	*fhuA2* [*lon*] *ompT gal sulA11* R(*mcr-73::miniTn10*--Tet^S)2 [*dcm*] R(*zgb-210::Tn10*--Tet^S) *endA1* Δ(*mcrC-mrr*)*114::IS10*

E. coli 5α F' and Express were obtained from New England Biolabs (Beverly, Massachusetts, USA), *E. coli* TOP10 F' and BLR from Invitrogen (Karlsruhe, Germany) and *E. coli* SURE from Stratagene (La Jolla, California, USA).

***Tetrahymena thermophila* strains**

Strain Name	Strain ID	Genetic Description
CU427.4	CU427.4	*chx1-1/chx1-1* (*CHX1*; cy-s, VI)
CU428.2	SD00178	*mpr1-1/mpr1-1* (*MPR1*; mp-s, VII)

All *Tetrahymena* strains were ordered from the *Tetrahymena* Stock Center at Cornell University (Ithaca, New York, USA; see also tetrahymena.vet.cornell.edu).

4.1.8 Oligonucleotides

The oligonucleotides used in this study were designed with Oligo 6.8 (Molecular Biology Insights Inc., Cascade, Colorado, USA) and synthesized by Sigma-Aldrich (St. Louis, Missouri, USA). Standard oligos were purified by desalting. HPLC-purified and 5'-phosphorylated oligos were used for mutagenesis and cloning with annealed oligos.

4.1.9 Gene Synthesis

In the scope of this work, FG repeats and FG repeat mutants from a range of diverse organisms were studied. The majority of the genes encoding these proteins were designed for optimized expression in *E. coli* using GeneDesigner 2 (DNA2.0, Menlo Park, California, USA), synthesized *de novo* by GenScript (Piscataway, New Jersey, USA) and subsequently cloned into the desired target expression vectors.

4.1.10 Plasmids

The plasmid generated and/or used in this study originate from either the high-copy plasmid pQE80 (ColE1 ori; Quiagen, Hilden, Germany), the low-copy plasmid pZA4 (p15A ori; kind gift of R. Lutz) or the single-copy, F plasmid-derived vector pSMART (OriV; Lucigen, Middleton, Wisconsin, USA), depending on the cloning and expression strategy followed for a given construct. All expression vectors are under the control of the *lac* promotor and contain the *lac*I gene encoding the *lac* repressor, which is required to prevent premature expression before induction. The constructs were generated as described in the Methods. Plasmids named pSF were kindly provided by Dr. Steffen Frey. An overview of the expression constructs generated and/or used in this study can be found in Table 4.1.

4.2 Methods

4.2.1 Bioinformatics Analysis of Nup98-Derived FG Repeats and Ran

Identification of Nup98 Homologs in Eukarya

The Nucleoporin2 domain from human Nup98 (GI: 56549643, see also Fontoura et al., 1999) was used as an initial bait for the systematic identification of Nup98 homologs in Eukarya. Note that in contrast to the FG repeat domain, it is the only folded part of Nup98, with crystal structures available (PDB entry: 1KO6, Hodel et al., 2002). As it is indeed also highly conserved, the Nucleoporin2 domain was considered a diagnostic feature for Nup98 (see also Section 2.1.1.). A database of Nup98 homologs was complied by searching the non-redundant NCBI protein database in every eukaryotic clade on an supergroup-level (as described by Simpson and Roger, 2004) with the amino acid sequence of the human Nucleoporin2 domain using the BLAST tool. Also, the top hits in each supergroup were used for fine searches within that respective taxonomical entity. In case the NCBI protein database was lacking proteome assemblies from key organisms, as for example in the case of the only completely sequenced member of the Eukaryotic clade Rhizaria, *Bigelowiella natans*, the online resources of the respective genome projects were resorted to. Additionally, the Pfam database entry for the Nucleoporin autopeptidase domain (http://pfam.sanger.ac.uk/family/nucleoporin2) was referred to for cross-checking and completing the custom Nup98 homolog database created in this study.

Extraction of FG Repeat Regions From Full-Length Nup98 Homologs

Unambiguous determination of the exact FG repeat region for every identified Nup98 homolog was not feasible, as the field currently lacks an unquestioned definition. In particular, the domain boundaries are disputed, mainly in regard as to how far from the two terminal FG motifs the domain extends. Notably, it was also shown that hydrophobic clusters other than canonical FG motifs can serve as NTRs binding sites as well (Labokha, A., personal communication), further complicating the issue.

In this study, FG repeat domains where identified semi-automatically with custom-written Mathematica routines and manual optimization. The program automatically extracted the 'core' FG repeat regions based on the presence of FG and related motifs (e.g. FA, FI, FS, FT, IG, LG, LL), thus essentially following a modified version of the FG repeat definition implemented by (Strawn et al., 2004). Clear 'outliers' were manually excluded after inspection of the suggested FG repeat region, thereby correcting the predictions. Note that all Nup98-derived FG repeat regions chosen for further biochemical analysis were revisited and redefined taking also secondary structure and disorder predictions, as well as FG repeat-typical local amino acid bias and resemblance to known FG repeat regions into consideration.

Basic Statistical Analysis of the Extracted FG Repeat Regions

The extracted FG repeat regions were analyzed in regard to their amino acid composition (i.e. the frequency of the 20 standard amino acids), their charge density and their overall hydrophobicity (as calculated according to Fauchere and Pliska, 1983). For all parameters, the average and standard deviation were calculated. For comparison, the same basic statistical analysis was performed for a collection of 43,634 globular proteins from the PDB. All analyses were performed with custom-written Mathematica routines.

To identify correlations between sequence features in the FG repeat dataset, a correlation matrix for the 27 parameters 'frequency of the twenty standard amino acids', 'NQ- and FILV-content', 'overall hydrophobicity', 'net charge' and 'FG motif occurrence' was calculated using the available in-build Mathematica algorithms in custom-written routines.

Cluster Analysis of the Nup98-Derived FG Repeat Database

To obtain a clearer picture of the similarities and differences between the Nup98-derived FG repeats, the dataset was further examined by cluster analysis, a computational procedure used to identify similar items in large datasets (Fayyad et al., 1996). Cluster analysis does not rely on a single algorithm, but rather can be performed in different ways. For example, datasets can be hierarchically clustered using divisive (top-down) or agglomerative (bottom-up) algorithms, depending on wether they part a complete dataset incrementally into smaller clusters, or merge multiple small clusters successively into larger ones, respectively. In such cases, the final number of clusters is determined by the algorithm. In contrast, datasets can also be clustered into a fixed number of subgroups. Note that in either case, the data is not clustered according into content-defined groups or in any other way interpreted during cluster analysis. *Post-hoc* evaluation and interpretation of the results depend entirely on the experimenter.

In this study, custom-written Mathematica programs based on the in-build cluster analysis algorithms and their associated options were used to sort the FG repeat database. The performance of different algorithms was compared primarily based on their ability to faithfully identify and segregate a control group consisting of folded globular proteins. Note that the most consistent and plausible results were obtained with the Euclidean-Distance algorithm operated in local optimization mode.

Identification of Ran Homologs in Eukarya

To compile a database of Ran homologs, human Ran (GI: 5453555, Bischoff and Ponstingl, 1991b) was used as a bait to search the non-redundant protein database at NCBI in multiple BLAST queries each limited to an individual eukaryotic supergroup, thereby systematically screening the entire eukaryotic biodiversity (see also above).

Analysis of the Ran Size Conservation

Custom-written Mathematica routines were used to calculate the molecular weight and amino acid sequence length for every entry of the entire Ran homolog database. The results were plotted supergroup-wise and statistically analyzed by a nonparametric one-way ANOVA test (after Kruskal and Wallis, 1952), followed by a Dunn's post-test to compare the column means) implemented in the Prism5 software. The sequence identities between the least conserved Ran homologs were determined based on a multiple sequence alignment of selected species (i.e. one representative from each supergroup) generated with ClustalW (Chenna et al., 2003) using the MegAlign tool of the Lasergene software collection. Structure-aided TCoffee multiple sequence alignments based on the RanGTP (PDB entry: 1RRP) and RanGDP (PDB entry: 3GJ0) structures were used to calculate a heat maps of sequence conservation with the ProtSkin web tool. The amino acids in the Ran structures were then colored according to the obtained heat maps using PyMol to create a graphical representation of the Ran conservation.

4.2.2 Standard Molecular Biology Techniques

DNA Purification from *E. Coli* Cultures

Propagated plasmid DNA was purified from late exponential to stationary phase *E. coli* o/n cultures on an analytical scale using the NucleoSpin Plasmid kit (Macherey Nagel, Düren, Germany) or on a preparative scale using the NucleoBond PC100 (Macherey Nagel, Düren, Germany), in either case according to the manufacturer's instructions.

Determination of DNA Concentrations

The concentrations of aqueous DNA solutions were determined based on the characteristic absorption of light by DNA at $\lambda = 260$nm (A_{260}), with $A_{260} = 1.0$

corresponding to a DNA concentration of 50 ng/µl (Sambrook, 2001). All measurements were done with a NanoDrop-1000 spectrophotometer.

Polymerase Chain Reaction

To amplify target DNA fragments from DNA templates, polymerase chain reactions (PCRs; Mullis et al., 1986) were performed with target-specific oligonucleotide primers. The primers were designed such that they also contained engineered restriction enzyme cleavage sites to allow for subsequent cloning of the PCR products. The high-fidelity, proof-reading and thermostable Pfu or PfuS polymerases (at 100ng/µl) were used in the amplification reactions for general cloning, both in combination with a thermostable Pab pyrophosphatase (15ng/µl) and Pab dUTPase (2.5ng/µl). For colony PCRs (see below), a Tth polymerase mix consisting of Tth polymerase (100ng/µl) and Pab pyrophosphatase (15ng/µl) was used. All enzymes were expressed and purified in our department. In general, 100µl PCR mixture contained:

x µl	template DNA (usually 50-100ng)
10 µl	10x dNTP-mix (containing 2.5mM of each dATP, dCTP, dGTP, dTTP)
1 µl	forward primer (100µM)
1 µl	reverse primer (100µM)
1 µl	DNA polymerase enzyme blend
10 µl	polymerase buffer (10x)
x µl	ddH$_2$0 (adjust to final volume)

Primarily to reduce secondary structure, 10µl of DMSO were additionally added to the reaction mixtures when required. The standard PCR protocol followed in this study, exemplified here for PfuS polymerase, included:

Initial denaturation:
98°C   2 min

30-35 amplification cycles:
98°C   30 sec
T_{anneal} 30 sec
68°C t_{ampl}

Final extension:
68°C   5 min
10°C   ∞

T_{anneal} represents the annealing temperature required for the specific primer pair used in the reaction, which was routinely calculated with the Oligo software package. t_{ampl} is the variable duration for the extension step, which depends on the amplification rate of the chosen polymerase (estimated to be ~30sec/kb for PfuS-, xsec/kb for the more Pfu- and ~40-60sec/kb for the Tth-polymerase) and the length of the target DNA fragment. Note that in reactions were Tth polymerase was employed, the denaturing temperature was lowered to 94°C and the annealing temperature raised to 72°C.

After the reaction, PCR products were purified using the Zymoclean Gel DNA Recovery kit (Zymo Research, Freiburg, Germany) according to the manufacturer's instructions, with the sole exception that an equal volume of 2-propanol was added to the PCR reactions prior to continuation with the provided protocol. Subsequently, the purified PCR products were digested with restriction enzymes (see below) to be cloned into desired target vectors.

Mutagenesis Polymerase Chain Reaction

In order to introduce specific mutations in a given gene, e.g. with the intend of changing single amino acids encoded in the gene, deleting small parts of the gene, or to insert additional base pairs that are restriction sites and/or encode for additional amino acids, mutagenesis PCR was routinely employed. For this purpose, standard PCRs (as described above) were performed with primers designed such that not only a small fragment, but the entire plasmid of interest was amplified. However, due to intended non-complementary base pairing of the primers to a specific locus of the gene, a desired mutation could be manifested in the amplified plasmid DNA. Note that only HPLC-purified and 5'-phosphorylated primers were used for mutagenesis PCR.

After the reaction, the purified PCR product was treated with the restriction enzyme DpnI to specifically digest the methylated template plasmid not bearing the desired mutation, and further purified via Agarose gel electrophoresis. Eventually, the amplified plasmid was ligated and transformed into appropriate *E. coli* strains for propagation (see below).

DNA Cleavage with Restriction Enzymes

All restriction endonucleases were obtained from New England Biolabs, Beverly, Massachusetts, USA. The cleavage reactions were performed according to the recommendations of the manufacturer, using the provided buffers and reagents. At least a two-fold excess of endonucleases was used for cleavage reactions, whereas the amount of enzyme was decreased for o/n reactions. In cases were vector DNA has been cleaved, fast alkaline phosphatase (Fermentas, St.Leon-Roth, Germany) was additionally added to the reactions at least 10 minutes prior to completion, allowing de-phosphorylation of the cleaved DNA and thus preventing re-ligation during subsequent cloning steps. Restriction enzymes where generally chosen to generate complementary overhangs in insert and vector DNA. Alternatively, blunt ends were required for some cloning strategies.

DNA Gel Electrophoresis

To size-separate the DNA fragments resulting from PCR amplification or restriction enzyme digestion, agarose gel electrophoresis was employed as described (Sambrook, 2001) with agarose concentrations varying from 0.8% (w/v) to 2.0% (w/v), depending on the expected sizes of the DNA fragments to be separated. DNA samples were combined with at least 1:10 (v/v) Orange-G loading buffer. As a reference, a standard marker with DNA fragments of defined sizes (DNA-ladder; Fermentas, St. Leon-Roth, Germany) was routinely included. To allow direct visualization of the separated DNA fragments in the agarose gels under UV light, ~0.05 µg/ml ethidium bromide was added to the liquid agarose prior to gelation. Agarose gels were formed and run in 1x TAE.

DNA Extraction from Agarose Gels

Size-separated DNA fragments were cut out of agarose gels using a scalpel and purified from the thus obtained gel slices using the Zymoclean Gel DNA Recovery kit (Zymo Research, Freiburg, Germany) according to the manufacturer's instructions.

Ligation of DNA

In the final step of cloning, appropriately cleaved inserts (~60 fmol or 40ng/kb) and vectors (~30 fmol or 20ng/kb) were ligated using T4 DNA ligase (100ng/µl; expressed and purified in our department). As a negative control and to assess the relative ligation efficiency, the digested and dephosphorylated vector DNA was incubated without insert. The ligation reactions were performed in 1x ligation buffer, supplemented with an energy-regenerating system (1x E-mix, see Materials) and 5% PEG-4000, at 37°C for 30 min or o/n at either RT or 16°C.

Transformation of DNA into Electro-competent *E. coli* Cells

For the preparation of electro-competent E. Coli cells, into which ligation reactions or purified plasmid DNA can be transformed by electroporation, a pre-culture of an appropriate strain (see Materials) was grown o/n at 37°C in liquid SOB medium containing the required antibiotics. The pre-culture was diluted 1:200 with pre-warmed SOB medium and further incubated at 37°C until the culture reached an optical density of

0.8 at 600nm (OD_{600} = 0.8). The bacteria were then pelleted at 3500 rpm for 10 min at 4°C in an F10 rotor (see Materials). The supernatant was removed carefully and and the pelleted cells were resuspended in approx. 100 ml 1.4% (v/v) Glycerol, only to be pelleted and washed with 1.4% (v/v) Glycerol again. After the final centrifugation step, the bacteria were eventually taken up in 14% (v/v) Glycerol, aliquoted, frozen in liquid nitrogen and stored at –80°C until further use. Note that electro-competent cells were prepared by G. Kopp for general use in the department.

For electroporation, either 0.5-1µl of a ligation reaction or ~50ng of plasmid DNA was mixed with 45µl of electro-competent *E. coli* cells on ice in an electroporation cuvette (BioRad, Hercules, California, USA) and subjected to electroporation using the GenePulser™ (BioRad, Hercules, California, USA) according to the manufacturer's instructions. Subsequently, the cells were recovered in 1ml 2YT medium supplemented with 0.2% Glucose, but without antibiotics, and incubated for 1 hour at 37°C while shaking. Finally, the bacteria were plated on selective agar plates containing the appropriate antibiotic to select for plasmid-containing cells and incubated o/n at 37°C.

Employed Screening Techniques for Positive Clones

In order to evaluate the success of a given cloning procedure, different techniques to screen for the correct clones were used. Generally, small-scale test expressions were used for screening expression constructs, as this procedure not only allows to identify positive clones, but also provides first information regarding the expression levels of the construct. Whenever test expression was not applicable, colony PCRs (see below) or analytical restriction enzyme digests were performed. Note that all identified clones were verified by DNA sequencing (see below).

Colony Polymerase Chain Reaction

Colony PCR is an analytical method used to amplify DNA directly from a (bacterial) cell culture in order to identify clones carrying the gene or plasmid of interest. For this purpose, single bacterial colonies were picked from agarose selection plates and resuspended in 50µl 2YT medium, of which 1µl was added to a 19µl PCR mixture as the template instead of purified DNA. Tth polymerase was chosen for colony PCRs. For analysis, agarose gel electrophoresis (see above) was employed. In case of a positive hit, the remaining cell suspension was used to inoculate a culture for plasmid DNA preparation (see above).

DNA Sequencing

All DNA sequencing reactions were performed by Seqlab Sequence Laboratories (Göttingen, Germany). The results were analyzed with the SeqMan program included in the Lasergene 8 software package (DNA-Star, Madison, Wisconsin, USA).

4.2.3 Protein Expression and Purification

Expression

All proteins were expressed in appropriate *E. coli* expression strains (see Materials). Notably, the quality and yield of a recombinantly expressed protein depends on a plethora of parameters ranging from factors relating to the expression construct itself (such as AT-content of the coding sequence, secondary structure of the corresponding transcribed RNA, included affinity or solubility-enhancing tags, etc.) to the expression conditions employed (i.e. the chosen expression strains, temperatures and times, as well as the induction strength). Ideal expression conditions were determined for each protein of interest and are summarized in Table 4.1.

In general, a plasmid encoding for a given recombinant protein of interest was transformed into an electro-competent *E. coli* expression strain and plated on selective agarose plates after recovery in 1ml 2YT medium plus Glucose. A single colony was picked the following day and used to inocculate a pre-culture, which was grown o/n at 30°C in either 2YT or TB medium supplemented with the appropriate antibiotic. The stationary phase pre-cultures were diluted by addition of fresh medium and incubated at the desired expression temperature until they reached an OD_- of 0.8-1.0 (in case of expression in 2YT medium) or 3.0-5.0 (in case of expression in TB medium). Protein expression was induced by the addition of 25µM-1mM Isopropyl-β-D-thiogalactopyranosid (IPTG), depending on the expression temperature and time. Note that protein expression was routinely performed either at 18°C for 14-16 h (for subsequent native purifications) or at 37°C for 4-5 h (for subsequent purifications under denaturing conditions). Prior to harvesting, 1mM phenylmethanesulfonylfluoride (PMSF) was added directly to o/n cultures intended for native protein purification, whereas 10mM ethylenediaminetetraacetic acid (EDTA) was added to the centrifugation buckets in every case, irrespectively of the purification protocol subsequently followed. The cultures were then harvested by centrifugation (10min, 4°C, 5000rpm) and resuspended to an OD_- of up to 200 in either native buffer or 8.5M GHCl (refer to Table 4.1 for details). Finally, the resuspended bacteria were snap-frozen in liquid nitrogen and stored at − 80°C, when otherwise not directly processed.

Cell Lysis

The bacterial cell walls were primarily disrupted due to mechanical stress during freeze/thawing. Only in case of native purifications, Lysozyme (1µg/ml) was occasionally added to the cell suspension prior to freezing, thereby further facilitating cell lysis by additional enzymatic hydrolysis of the peptidoglycan layer of the bacterial cell wall. 2-10mM DTT were added after thawing to prevent oxidation of proteins. To reduce the viscosity of the thus obtained crude lysates, the released DNA was shorn apart by sonication (4x 1min on ice, 50% duty cycle, max. output). To remove cell debris and large aggregates, the crude lysates were ultracentrifuged for at least 1.5h in a T647.5 or T1250 rotor at 38,000rpm, either at 4°C (native purifications) or at 20°C (denaturing conditions). The supernatant was carefully taken off, yielding the cleared lysate that served as the starting material for protein purification.

Native Protein Purification

Purification Step 1: Native Ni$^+$ Affinity Chromatography

To separate the recombinantly expressed proteins of interest from the bulk of the cleared lysates, Ni^{2+} affinity chromatography was used. This purification technique relies on the ability of the Imidazole ring of the Histidine side chain to complex bivalent metal ions (e.g. Ni^{2+}, Co^{2+} or Cu^{2+}) at physiological pH. Thus, by immobilizing the metal ions on solid supports such as porous silica beads, affinity matrices can be generated, to which recombinant target proteins 'tagged' with a stretch of poly-Histidine residues can be bound and thereby specifically 'pulled' out of complex protein mixtures such as whole-cell lysates. For this purpose, expression constructs in this study contained either standard His10 or His14 tags, as well as newly designed His tags optimized for protein expression, purification and hydrogel formation (see also Table 4.1).

For native purifications, an appropriate volume of 24% Ni^{2+} EDTA 500Å silica matrix (custom-made by Dr. Dirk Görlich) was equilibrated with the respective native resuspension buffer used and added to the cleared lysate. The His-tagged target protein was allowed to bind for at least 1h at 4°C while rotating. Low concentrations of imidazole (i.e. 10-25mM) were included to reduce unspecific binding of bacterial proteins. The mixture was then applied to a gravity flow column (Sigma-Aldrich, St. Louis, Missouri, USA), allowing the matrix to settle. The flow-through was collected and the matrix bed washed thoroughly with resuspension buffer. Elution from the affinity matrix was subsequently achieved by competitive binding of excess Imidazole in the elution buffer (resuspension buffer supplemented with 500mM Imidazole), which was applied gradually in 100-500µl steps. The elution fractions containing the highest protein concentration were identified by amido black staining of 1µl samples spotted onto a nitrocellulose membrane or visual inspection of fluorescence intensity (in case of fluorescently-labeled proteins), and then pooled. In case the combined elution fraction were not directly processed further, they were complemented with 250mM Sucrose to prevent freezing damage during subsequent snap-freezing in liquid nitrogen and storage at – 80°C. Prior to freezing, protein concentrations were determined spectroscopically by absorption measurements at 280nm and the characteristic absorption wavelengths of the fluorescent tags, when applicable. All purification steps were followed by SDS-PAGE analysis.

Endoprotease Cleavage to Specifically Remove Tags

Besides the poly-Histidine stretches used for initial purification from whole-cell bacterial lysates, other tags/fusion partners with beneficial properties were required in some cases, e.g. to enhance the expression levels or solubility of the recombinant target proteins (see also Table 4.1). However, such tags can be rather bulky and disadvantageous for downstream applications. To separate larger tags from target proteins, defined cleavage sites for specific endoproteases can be included in the expression constructs. For this purpose, either the TEV protease recognition sequence or small ubiquitin-like modifier (SUMO) protein were fused in-between the N-terminal tag and the target protein in this study.

For TEV fusions, cleavage was performed by addition of TEV protease (expressed and purified in our lab by Dr. Steffen Frey) directly to the Ni^{2+} eluate. The cleavage reactions were incubated either for 1h at RT or o/n at 4°C. A 100-fold molar excess of substrate over protease was used. Afterwards, the target protein was further purified via gel filtration and/or reverse Ni^{2+} affinity chromatography (see below).

SUMOylation is a common post-translational protein modification involved in various cellular processes, in which a SUMO protein is covalently attached to a target protein. The modification is reversible, as specific proteases can cleave off the SUMO moiety again. As a tool in recombinant protein technology, SUMO is genetically fused to a target protein and serves as an expression level and solubility enhancer, as well as recognition unit for the specific removal of itself (and everything additionally fused to it, such as e.g. a poly-His tag) from the target protein. SUMO tags were specifically cleaved off using SUMO protease in a 1:10,000 molar ratio of protease to substrate. Reactions were carried out for 1h at 4°C. The target protein was either further purified via gel filtration or alternatively separated from the His-tagged SUMO moiety via reverse Ni^{2+} affinity chromatography (see below).

Purification Step 2: Gel Filtration Chromatography

Gel filtration is a chromatography technique used to separate molecules according to differences in size and conformation as they pass through a porous medium packed into a column. It is based on the principle that smaller molecules can efficiently enter the pores of the medium, whereas larger molecules are excluded and pass alongside the beads, hence eluting first. In this study, Superdex™ columns (Pharmacia, Upsala, Sweden), a spherical composite of cross-linked agarose and dextran, were used. All gel filtrations were performed using an Äkta Purifier (Pharmacia, Upsala, Sweden) setup.

Prior to use, the gel filtration column was equilibrated with 1.5-column volumes of GF buffer (20mM Tris pH7.5, 130mM NaCl, 2mM MgCl, 2mM DTT) using the default flow rate. To remove precipitates, protein samples were ultracentrifuged in an appropriate rotor (see Materials) and only then injected onto the equilibrated column. The gel filtration runs were performed according to the column specifications provided by the manufacturer. All elution fractions were collected and analyzed with SDS-PAGE. The fractions containing the pure target protein were pooled and complemented with 250mM Sucrose. After protein concentration determination via absorbance at 280nm (and at the characteristic absorption wavelength of the attached fluorophore, when applicable), the sample was aliquoted, snap-frozen in liquid nitrogen and stored at –80°C.

Purification Step 3: Reverse Ni-Affinity Chromatography

To remove possible contaminations, primarily un-cleaved target protein and the released tags, reverse Ni^{2+} affinity chromatography was performed. In contrast to conventional Ni^{2+} affinity chromatography, only the aforementioned contaminations will bind to the affinity matrix in this case, whereas the untagged target protein will be present in pure form in the flow-through. An equal amount of Ni^{2+} EDTA 500Å silica matrix as taken for the initial affinity purification step was equilibrated with GF buffer (20mM Tris pH7.5, 130mM NaCl, 2mM $MgCl_2$, 2mM DTT) and utilized to filter out remaining His-tagged contaminations. Samples from the flow-through, as well as from the elution fractions after reverse Ni^{2+} affinity chromatography were analyzed with SDS-PAGE for quality control.

Note that reverse Ni^{2+} affinity chromatography does not work when the applied sample still contains high residual concentrations of Imidazole from the initial affinity purification step. Thus, for samples that have not been previously purified via gel filtration chromatography, a buffer exchange was performed prior to reverse Ni^{2+} affinity chromatography. For this purpose, PD10 desalting columns (Amersham Biosciences, Uppsala, Sweden) were first equilibrated with an appropriate buffer (lacking Imidazole). Then, the Imidazole-containing sample was added and allowed to enter the column. Finally, the sample was eluted in an Imidazole-free form with equilibration buffer. All steps were performed according to the manufacturer's instructions.

Protein Purification Under Denaturing Conditions

Purification Step 1: Ni-Affinity Chromatography Under Denaturing Conditions

For denaturing purifications, the cleared bacterial lysate was first buffered with 200mM Tris pH8.0, supplemented with 5-15mM Imidazole (depending on the His-tag of the construct to be purified; see also Table 4.1) and adjusted to 10mM DTT. The buffered lysate was then incubated o/n at RT with 50% Ni^{2+} EDTA 1000Å silica matrix (custom-made by Dr. Dirk Görlich) under an Argon layer. Note that the matrix was previously thoroughly washed and equilibrated with ddH_2O and denaturing equilibration buffer (100mM Tris pH8.0, 6.0M GHCl, 5mM Imidazole), respectively. After incubation, the mixture was applied to a gravity flow chromatography column (Sigma-Aldrich, St. Louis, Missouri, USA) and the matrix allowed to settle, while the flow-through was collected. The matrix bed wash then washed thoroughly with denaturing wash buffer (200mM Tris pH8.3, 8.0M GHCl, 5-15mM Imidazole (see above), 10mM DTT). For elution, an elution buffer containing 200mM Tris pH8.0, 8.0M GHCl, 500mM Imidazole and 10mM DTT was used and applied stepwise to the matrix bed, thereby preventing unnecessary dilution of the sample. 1µl of each elution fraction was spotted onto a nitrocellulose membrane and stained with amido black to identify the protein-containing fractions, which were subsequently pooled. Samples for SDS-PAGE analysis were taken from the input, flow-through and elution fractions. Note that precipitation with 90% 2-propanol was required prior to uptake of the samples in SDS sample buffer, as GHCl and SDS form insoluble aggregates that would obscure the electrophoretic analysis. The pooled elution fractions were snap-frozen in liquid nitrogen and stored at –80°C until further use.

Purification Step 2: Reverse-Phase HPLC

The pooled elution fractions from denaturing Ni affinity chromatography were further purified using reverse-phase high-performance liquid chromatography (HPLC) on a C18 column (Vydac 218 TP 10/22) in an industry-scale Jesco PU-2087 setup with appropriate detection systems (Jesco, Essex, United Kingdom). In this chromatography method, proteins are first bound from an aqueous buffer phase to the n-alkyl hydrocarbon column via hydrophobic interactions, and then eluted with a gradient of organic solvent. Depending on their individual constitution, i.e. in particular their hydrophobic character, some proteins will elute earlier than others, thus allowing their separation and purification. Here, increasing concentrations of acetonitrile in 0.15% TFA were used. The quality of all reverse-phase HPLC runs was controlled with SDS-PAGE. For this purpose, samples from the gradient elution fractions were taken and assimilated in SDS sample buffer after solvent evaporation using a SpeedVac (RVC 2-18 series; Christ, Osterode, Germany). Please note that all reverse-phase HPLC runs were operated and supervised by Jürgen Schünemann.

Lyophilization of Purified Proteins

After final reverse-phase HPLC purification, the proteins were snap-frozen in liquid nitrogen and freeze-dried using an Alpha 1-2 LD plus bench-top lyophilizer (Christ, Osterode, Germany), allowing the solvent to completely sublimate. The dried proteins were stored at –80°C until further use.

Table 4.1. *Proteins expressed or used in this study.*

– FG Repeat Domains and Mutants –						
Protein Name	Purified Protein	Organism of Origin	Plasmid	Expression Conditions	Purification Conditions	Reference
TbNup64	H14-TEV-TbNup64-Cys	*Trypanosoma brucei*	pHBS301			This study
TbNup75	H14-TbNup75	*Trypanosoma brucei*	pHBS302			This study
TbNup140	H14-TEV-TbNup140-Cys	*Trypanosoma brucei*	pHBS297			This study
TbNup158	H14-TEV-TbNup158-Cys	*Trypanosoma brucei*	pHBS249			This study
GlNup45	H14-TEV-GlNup45-Cys	*Giardia lamblia*	pHBS239			This study
BnNup98	HKx18-BnNup98-Cys	*Bigelowiella natanas*	pHBS502			This study
TtNup72	HKx18-TtNup72-Cys	*Tetrahymena thermophila*	pHBS310			This study
TtMacNup98A	HQx18-TtMacNup98A-Cys	*Tetrahymena thermophila*	pHBS418			This study
TtMacNup98A-FSFG$_{x18}$	HQx18-TtMacNup98A-FxFG$_{x18}$-Cys	Hybrid construct	pHBS487			This study
TtMacNup98A GLFG-NIFN	H14-TEV-TtMacNup98A GFLG-NIFN-Cys	Synthetic construct	pHBS135	*E. coli* BLR or Express, 1mM IPTG @OD$_{600}$≈3-5, 4h, 37°C	Denaturing protein purification	This study
TtMacNup98B	HQx18-TtMacNup98B-Cys	*Tetrahymena thermophila*	pHBS419			This study
TtMacNup98B GLFG-NIFN	H14-TEV-TtMacNup98B GFLG-NIFN-Cys	Synthetic construct	pHBS137			This study
TtMicNup98A	HQx18-TtMicNup98A-Cys	*Tetrahymena thermophila*	pHBS318			This study
TtMicNup98A-FSFG$_{x18}$	HQx18-TtMicNup98A-FxFG$_{x18}$-Cys	Hybrid construct	pHBS489			This study
MpNup84	H14-TEV-MpNup84-Cys	*Micromonas pusilla*	pHBS240			This study
AtNup98A	H14-TEV-AtNup98A-Cys	*Arabidopsis thaliana*	pHBS382			This study
AtNup98B	H14-TEV-AtNup98B-Cys	*Arabidopsis thaliana*	pHBS383			This study
DdNup220	H14-TEV-DdNup220-Cys	*Dictyostelium discoideum*	pHBS241			This study
ScNup145	HQx18-ScNup145	*Saccharomyces cerevisiae*	pHBS492			This study
CeNup98	HKx18-CeNup98-Cys	*Caenorhabditis elegans*	pHBS504			This study
DmNup98	HKx18-DmNup98-Cys	*Drsosophila melongaster*	pHBS03			This study
BfNup98	HKx18-BfNup98-Cys	*Branchiostoma floridae*	pHBS505			This study
XlNup98	H14-TEV-XlNup98$_{2-484}$-Cys	*Xenopus laevis*				(Hülsmann et al., 2012)

Protein Name	Purified Protein	Organism of Origin	Plasmid	Expression Conditions	Purification Conditions	Reference
HsNup98	HQx18-HsNup98	*Homo sapiens*	pHBS491			This study; A. Labohka, unpublished
Nsp1$_{2-601}$	H14-TEV-Nsp1$_{2-601}$-Cys	*Saccharomyces cerevisiae*	pSF1020			(Frey and Görlich, 2007)
Nsp1$_{1-175}$	H14-TEV-Nsp1$_{2-175}$-Cys	*Saccharomyces cerevisiae*	pSF902			This study, (Ader et al., 2010)
Nsp1$_{2-277}$	H14-TEV-Nsp1$_{2-277}$-Cys	*Saccharomyces cerevisiae*	pSF1074			S. Frey, unpublished
Nsp1$_{2-277}$-GFP	H14-TEV-Nsp1$_{2-277}$-shGFP2	Synthetic construct	pHBS468		Native protein purification (50mM Tris pH7.5, 500mM NaCl)	This study
Nsp1 WT fragement	HKx18-Nsp1 WT-Cys	Synthetic construct	pHBS449			This study
Nsp1 Mut1 fragement	HKx18-Nsp1 Mut1-Cys	Synthetic construct	pHBS441			This study
Nsp1 Mut2 fragement	HKx18-Nsp1 Mut2-Cys	Synthetic construct	pHBS442			This study
Nsp1 Mut3 fragement	HKx18-Nsp1 Mut3-Cys	Synthetic construct	pHBS443		Native protein purification (44mM Tris pH7.5, 290mM NaCl, 4.4mM MgCl$_2$)	This study
Nsp1 Mut4 fragement	HKx18-Nsp1 Mut4-Cys	Synthetic construct	pHBS444			This study
Nsp1 Mut5 fragement	HKx18-Nsp1 Mut5-Cys	Synthetic construct	pHBS445			This study
Nsp1 Mut6 fragement	HKx18-Nsp1 Mut6-Cys	Synthetic construct	pHBS446			This study
Nsp1 Mut7 fragement	HKx18-Nsp1 Mut7-Cys	Synthetic construct	pHBS447			This study
Nsp1 Mut8 fragement	HKx18-Nsp1 Mut8-Cys	Synthetic construct	pHBS448			This study
Nsp1$_{274-601}$	H10-TEV-Nsp1$_{274-601}$	*Saccharomyces cerevisiae*	pSF654			This study, (Ader et al., 2010)
Nsp1$_{274-601}$^{Mut8}	HKx18-Mut8-Cys	Hybrid construct	pHBS501			This study
Nsp1$_{2-175}$-Nsp1$_{274-601}^{Mut8}$	H14-TEV-Nsp1$_{2-175}$-Mut8-Cys	Hybrid construct	pHBS496		Denaturing protein purification	This study
Nsp1$_{2-175}$-SxSG$_{x16}$	H14-TEV-Nsp1$_{2-175}$-SxSG$_{x16}$	Hybrid construct	pHBS90			This study
Nsp1$_{2-175}$-Sp420	H14-TEV-Nsp1$_{2-175}$-Sp420-Cys	Hybrid construct	pHBS464			This study

– Amyloid Domains and Mutants –						
Protein Name	Purified Protein	Organism of Origin	Plasmid	Expression Conditions	Purification Conditions	Reference
Sup35 WT PrD	H14-TEV-Sup35$_{2-140}$-Cys	*Saccharomyces cerevisiae*	pHBS83	*E. coli* BLR, 1mM IPTG @OD$_{600}$≈3-5, 4h, 37°C	Denaturing protein purification	This study, (Ader et al., 2010)
Sup35 Y-F PrD	H14-TEV-Sup35$_{2-140}$ Y-F-Cys	*Saccharomyces cerevisiae*	pHBS84		Denaturing protein purification	This study

Protein Name	Purified Protein	Organism of Origin	Plasmid	Expression Conditions	Purification Conditions	Reference
Sup35 Y-FG PrD	H14-TEV-Sup35$_{2\text{-}140}$ Y-FG-Cys	*Saccharomyces cerevisiae*	pHBS85			This study
Sup35 WT PrD-Nsp1$_{176\text{-}601}$	H14-TEV-Sup35$_{2\text{-}140}$-Nsp1$_{176\text{-}601}$-Cys	Hybrid construct	pHBS132			This study
Sup35 Y-FG PrD-Nsp1$_{176\text{-}601}$	H14-TEV-Sup35$_{2\text{-}140}$ Y-FG-Nsp1$_{176\text{-}601}$-Cys	Hybrid construct	pHBS185			This study

– Nuclear Transport Receptors (NTRs), Import Adaptors, Cargoes –						
Protein Name	**Purified Protein**	**Organism of Origin**	**Plasmid**	**Expression Conditions**	**Purification Conditions**	**Reference**
ScImpβ (Kap95)	MRGS-Kap95-H6	*Saccharomyces cerevisiae*	pSF595	*E. coli* BLR, 0.2mM IPTG @OD$_{600}$≈1-2, o/n, 18°C	Native protein purification (44mM Tris pH7.5, 290mM NaCl, 4.4mM MgCl$_2$)	S. Frey, unpublished
HsImpβ	HsImpβ	*Homo sapiens*				I. Chaudhuri, unpublished
HsImpβ$_{45\text{-}462}$$^{\Delta\text{acidic loop}}$	HsImpβ$_{45\text{-}462}$$^{\Delta\text{acidic loop}}$-biotin	*Homo sapiens*			Native protein purification (50mM Tris pH7.5, 500mM NaCl)	T. Pleiner, unpublished
scIBB-MBP-GFP	scIBB-MBP-GFP	Hybrid construct	pSF851	*E. coli* BLR, 0.2mM IPTG @OD$_{600}$≈1-2, o/n, 18°C	Native protein purification (44mM Tris pH7.5, 290mM NaCl, 4.4mM MgCl$_2$)	This study
Rch1-IBB-MBP-GFP	Rch1-IBB-MBP-GFP	Hybrid construct				This study
Srp1	Srp1-GFP-Cys	*Saccharomyces cerevisiae*	pSF812			S. Frey, unpublished
mCherry	mCherry-Cys	Synthetic construct	pSF779		Native protein purification (44mM Tris pH7.5, 290mM NaCl, 4.4mM MgCl$_2$)	This study
MBP-mCherry	MBP-mCherry	Hybrid construct	pSF844			(Frey and Görlich, 2009)
tCherry	tCherry	Synthetic construct	pSF931			S. Frey, unpublished
TtImpB3	TtImpB3-mCherry/ TtImpB3-GFP	*Tetrahymena thermophila*	pHBS352/357	*E. coli* BLR, 0.2mM IPTG @OD$_{600}$≈1-2, o/n, 18°C	Native protein purification (50mM Tris pH7.5, 500mM NaCl)	This study
TtImpB6	TtImpB6-mCherry/ TtImpB6-GFP	*Tetrahymena thermophila*	pHBS353/358			This study
TtCAS	TtCAS/ GFP-TtCAS	*Tetrahymena thermophila*	pHBS340/328			This study
TtRanGTP	TtRanQ70L	*Tetrahymena thermophila*	pHBS290	*E. coli* Express, 0.2mM IPTG @OD$_{600}$≈5, o/n, 18°C	Native protein purification (50mM Na-Phosphate pH7.5, 500mM NaCl, 5mM MgAc)	This study

TtIMA1	ZZ-TEV-Sp-TtIMA1-H10/ TtIMA1-GFP	*Tetrahymena thermophila*	pHBS285/322			This study
TtIMA8	ZZ-TEV-Sp-TtIMA8-H10/ TtIMA8-GFP	*Tetrahymena thermophila*	pHBS286/323			This study
TtIMA12	ZZ-TEV-Sp-TtIMA12-H10/ TtIMA12-GFP	*Tetrahymena thermophila*	pHBS287/324			This study
TtIMA1$_{1-55}$	ZZ-TEV-Sp- TtIMA1$_{1-55}$-H10	Synthetic construct	pHBS366			This study
TtIMA1$_{54-526}$	ZZ-TEV-Sp- TtIMA1$_{54-526}$-H10	Synthetic construct	pHBS362	*E. coli* Express, 0.2mM IPTG @OD$_{600}$≈2-3, o/n, 18°C	Native protein purification (50mM Tris pH7.5, 500mM NaCl)	This study
TtIMA8$_{1-81}$	ZZ-TEV-Sp- TtIMA8$_{1-81}$-H10	Synthetic construct	pHBS367			This study
TtIMA8$_{1-97}$	ZZ-TEV-Sp- TtIMA8$_{1-97}$-H10	Synthetic construct	pHBS368			This study
TtIMA8$_{82-606}$	ZZ-TEV-Sp- TtIMA8$_{82-606}$-H10	Synthetic construct	pHBS363			This study
TtIMA8$_{98-606}$	ZZ-TEV-Sp- TtIMA8$_{98-606}$-H10	Synthetic construct	pHBS364			This study
TtIMA12$_{1-47}$	ZZ-TEV-Sp- TtIMA12$_{1-47}$-H10	Synthetic construct	pHBS369			This study
TtIMA12$_{48-530}$	ZZ-TEV-Sp- TtIMA12$_{48-530}$-H10	Synthetic construct	pHBS365			This study
TtH1	ZZ-TEV-Sp- TtH1-H10	*Tetrahymena thermophila*	pHBS379			This study
TtH2B	ZZ-TEV-Sp- TtH2B -H10	*Tetrahymena thermophila*	pHBS380	*E. coli* BLR, 0.2mM IPTG @OD$_{600}$≈2-3, o/n, 18°C	Denaturing protein purification with final native wash and elution step (44mM Tris pH7.5, 290mM NaCl, 4.4mM MgCl$_2$)	This study
TtMLH_β	ZZ-TEV-Sp- TtMLH_β -H10	*Tetrahymena thermophila*	pHBS397			This study
TtMLH_δ	ZZ-TEV-Sp- TtMLH_δ -H10	*Tetrahymena thermophila*	pHBS399			This study

4.2.4 SDS-PAGE

Analysis of protein samples was performed with discontinuous sodiumdodecylsulfate-polyacrylamide gel electrophoresis (SDS-PAGE) under denaturing conditions (Laemmli, 1970), followed by Coomassie staining. The composition of the gels run in our lab is as follows:

	Stacking Gel (4.5%) *(100ml)*	'Light' Gel (7.5%) *(200ml)*	'Heavy' Gel (16%) *(200ml)*
2M Tris pH8.8	—	40ml	40ml
0.5M Tris pH6.8	15ml	—	—
ddH2O	68ml	107ml	32ml
2M Sucrose	—	—	10ml
Glycerol (87%)	—	—	8ml
10% SDS	1ml	2ml	2ml
Rotiphorese Gel 30	15ml	51ml	108ml
TEMED	100µl	120µl	120µl
APS (10%)	1ml	2x 580µl	2x 580µl

Prior to loading, an appropriate volume of SDS sample buffer (see Materials) was added to the protein samples (max. 1:1 (v/v)). In case of highly cohesive proteins, a modified SDS sample buffer containing 6M Urea instead of Sucrose was used. Gels were run for >1 hour at 50mA constant current. Afterwards, proteins were fixed by heating the gel in 3% acetic acid and stained with an approximately 1:100 dilution of the Coomassie stock solution described in the Materials section. Gels were destained in ddH$_2$O and scanned using a Perfection V700 photo scanner (Epson, Tokyo, Japan). In case of subsequent MS/MS analysis, gels were additionally stained with colloidal Coomasssie (see Materials) prior to the excision of protein bands.

4.2.5 Protein Coupling to Maleimide-Sepharose

To create low-background affinity-matrices for pulldown experiments and binding assays, FG repeats and Sup35 PrD variants were covalently coupled to custom-made Maleimide Sepharose (by Dr. Dirk Görlich) via engineered C-terminal Cys-residues, resulting in stable carbon-sulfate bonds. For this purpose, 5mg of the lyophilized proteins of interest were resuspended in 5M GHCl (pH ~2.5 due to remaining TFA) and treated with Silica-bound Tris(2-carboxyethyl)phosphine (TCEP) (prepared by Dr. Dirk Görlich) for 2h at RT to completely reduce the thiol groups required for coupling. The reducing agent was easily removed by sedimenting the silica beads at 1000rpm for 1min in a FA-45 rotor (Eppendorf, Hamburg, Germany). The protein-containing supernatant was gently taken off and the availability of reduced thiol groups was verified with Ellmann's reagent (10mM DTNB in 1M Imidazole, pH8.0). For a standard coupling reaction, 1ml of Maleimide Sepharose was equilibrated with 1M GHCl prior to being combined with the reduced proteins of interest (i.e. 5mg protein/1ml matrix). To start the coupling reaction, 100mM K_2HPO_4 was added, resulting in a pH of ~7.5. All coupling reactions were performed o/n at RT while rotating. A lack of free thiol groups thereafter verified successful and complete coupling. To quench the remaining Maleimide groups and to passivate the Sepharose matrix, 0.25% (v/v) Mercaptoglycerol was added and the reactions were further incubated for 2h at RT while rotating. Afterwards, the affinity-matrices were successively washed with 5M GHCl, 1M GHCl and PBS, and finally stored in 30% degassed ethanol at 4°C.

4.2.6 Labeling of FG Repeats and Sup35 Prion Domains

The engineered cysteine residue at the C-termini of the FG repeats and Sup35 constructs was also used to generate fluorescent tracer molecules allowing the visualization of the hydrogels under the microscope in the setup described in Figure 2.5. Prior to labeling, the desired amounts of lyophilized FG repeats or Sup35 PrDs were solubilized in denaturing labeling buffer (i.e. 6M urea, 500mM NaCl and 0.5mM EDTA, buffered with potassium phosphate at pH7.0) and reduced with matrix-bound TCEP for 2h at RT (see above). Immediately afterwards, the supernatant containing the reduced protein was mixed with Atto647N-Maleimide (Sigma-Aldrich, Buchs, Switzerland) in a min. 1:1.5 molar ratio and incubated for 4-5 hours at RT. To stop the reaction and quench the unreacted Maleimide groups of the dye, 10mM DTT were added and the labeling reactions incubated for another 30-60 minutes at room temperature. The labelled protein was then precipitated with 1:10 (v/v) 2-propanol and resuspended in 6.5M GHCl. For final purification, the labeled proteins were purified via HPLC, snap-frozen in liquid nitrogen and lyophilized (see above). For final use, the tracers were re-suspended in 0.2% TFA and added during gel formation (see below) in concentrations ranging from 0.2-1µM.

4.2.7 Labeling of PEGs

Different-sized thiol-activated polyethylene glycol (PEG) molecules were ordered from Nanocs (Boston, Massachuchetts, USA) and labeled with Atto488-Maleimide (Sigma-Aldrich, Buchs, Switzerland) to obtain fluorescent probes for characterizing the hydrogel barrier tightness. For this purpose, 20kDa, 10kDa, 5kDa and 1kDa mPEG-thiols were resuspended in PBS and reduced with silica-bound TCEP (see above) for 2h at RT while rotating. The reducing agent was separated from the PEGs by centrifugation at 1000rpm in a table-top centrifuge (Eppendorf, Hamburg, Germany). All PEG-containing supernatants were carefully taken off, combined and mixed with Atto488-Maleimide in a 1:1.5 molar ratio, as recommended by the manufacturer. The coupling reaction was performed for 2h at RT while rotating. Afterwards, 20mM DTT was added to quench possibly remaining reactive groups.

In order to separate the labeled PEGs according to their size (and conformation), gel filtration chromatography (see above) on a Superdex™ 200 column equilibrated and run with 20mM Tris pH7.5, 130mM NaCl, 2mM MgCl was performed. All peaks were collected and analyzed by dynamic light scattering to determine their sizes. The size-separated samples were aliquoted, snap-frozen in liquid nitrogen and stored at –80°C until further use.

4.2.8 Preparation of FG Repeat Hydrogels

FG repeat hydrogels were formed as reported (Frey and Görlich, 2007; 2009). In summary, lyophilized FG repeats were dissolved at the indicated concentrations in 0.2% (v/v) TFA (in ddH$_2$O) or ddH$_2$O containing 0.25-1.0µM Atto647N-labelled Nsp1$_{2\text{-}175}^{F\text{-}I}$ or Sup35 WT PrD as tracers to visualize the hydrogels in the microscope. Approximately one microliter drops of the resuspended protein solutions were spotted onto uncoated 18-well µ-slides (ibidi, Martinsried, Germany) immediately after rehydration and allowed to gelate for at least 24h in a humid environment at RT. When required, the gelation process was delayed by including 0.5-2M GHCl during resuspension.

4.2.9 Hydrogel Permeation Assays

FG repeat hydrogels were equilibrated o/n in a large excess of buffer containing 20mM Tris-HCl pH7.5, 130mM NaCl, and 2mM MgOAc. Import of fluorescent substrate molecules into the FG repeat hydrogels was followed using an SP5 confocal laser scanning microscope and a 63x immersion objective (both from Leica, Mannheim, Germany). Essentially, the assays were performed as previously described (Frey and Görlich, 2007; 2009). In brief, the buffer-gel-boundary was placed at the center of the observed area and the focal plane was set to 5µm above the µ-slide surface. The 633nm laser line was used to excite the tracer molecules and thus monitor and adjust the position of the gel.

During data collection, frames of 2048 x 256 pixels were recorded every 60 sec for total of 30 minutes. The import substrates (standard concentrations used are 1µM fluorescently labelled-cargoes and 1.5µM NTRs) and controls (usually 3µM) were added after collection of the first frame. The 496nm and 561nm laser lines were used to excite GFP- and Cherry-tagged substrate or control molecules, respectively.

4.2.10 Analysis of the Hydrogel Permeation Assay Raw Data

The raw data obtained from imaging was analyzed with custom-written Mathematica routines (see also Frey and Görlich, 2007). Briefly, each individual frame was first converted into a concentration profile. In this process, all 256 lines of a frame were aligned to each other and then averaged to correct for deviations of the buffer-gel boundary from a perfectly perpendicular orientation. The aligned frames of each channel were once more aligned to calculate the multi-plots e.g. shown in Figure 2.5. Note that the image of the hydrogel served as reference during the latter alignment step. The corrected profiles were smoothed by an exponential filter.

4.2.11 Binding Assays

In the present study, binding assays were performed to test (i) the affinity of FG repeats and FG repeat mutants for NTRs, import adaptors or other FG repeats and (ii) the affinity of NTRs for cargoes. For these purposes, the respective baits were either covalently coupled to a Sepharose matrix (as described above) or immobilized via suitable tags to affinity matrices, as specified in the figure legends. In general, either His- or ZZ-tags and Ni^{2+}- or IgG-matrices were used, respectively. Please note that in all applicable cases, the IgG-Sepharose was pre-eluted with glycine buffer (pH2.2) prior to equilibration in the respective binding buffer.

Unless covalently coupled, His-tagged FG repeats were immobilized to Ni^{2+}-silica o/n at RT under denaturing conditions (200mM Tris pH8.0, 6.8M GHCl, 5mM imidazole, 10mM DTT). Prior to the binding assay, the repeats were washed with denaturing buffer containing gradually less GHCl to remove unbound material and stepwise reduce the GHCl concentration. Ultimately, the assay was performed in 50mM Tris pH7.5, 100mM NaCl, 2mM $MgCl_2$, 1% PEG4000 and 5mM DTT. Details on the type and concentrations of the preys used can be found in the figure legends. In case the binding assays were quantified, the matrix was gently pelleted after the indicated incubation times and the fluorescence in the supernatant measured with a NanoDrop spectrophotometer (Peqlab Biotechnologies GmbH, Erlangen, Germany). For elution, SDS sample buffer supplemented with 500mM imidazole was used.

Natively purified baits (i.e. ZZ-tagged TtIMAs) were incubated together with their preys (i.e. TtCAS in the presence or absence of TtRanQ70L) in a bacterial lysate adjusted to 50mM Tris pH7.5, 100mM NaCl, 2mM $MgCl_2$, 0.005% digitonin, 1x E-mix and 2mM DTT. Elution was performed with 1-2M $MgCl_2$. SDS-PAGE was used to analyze all binding assays performed in this study.

4.2.12 High-Cell Density Fermentation of *Tetrahymena thermophila*

All fermentations were carried out in a 13-liter Labfors-3 bench-top bioreactor (Infors, Bottmingen, Switzerland) with SSP medium (see Materials) in batch mode. Pre-cultures of *Tetrahymena thermophila* strain CU428.2 were grown in SSP medium at 30°C to a density of ~150,000 cells/ml, of which ~100 Mio. cells were used to inoculate 10-liter fermentation cultures. During fermentation, temperature and pH were kept constantly at 30°C and 7.0, respectively. To ensure proper aeration, the culture was stirred with a paddle impeller at max. 175 rpm and additionally infused with compressed air mixed with 22% O₂ at a constant gas flow rate of 1g/l, thereby maintaining the dissolved oxygen concentration at ~50% air saturation for all experiments. The cells were fermented until they reached a final density of ~1 Mio. cells/ml.

Prior to harvesting, 1mM PMSF and 10mM EDTA were added to the culture. Cells were pelleted at 4000rpm for 10min at 4°C in an F9 rotor, washed once in RES buffer (50mM HEPES pH7.6, 200mM KCl, 10mM MgCl2, 5mM EDTA, 250mM Sucrose, 2mM DTT), and resuspended to a final concentration of ~50 Mio. cells/ml in RES buffer. Note that 100µM of the Cysteine-protease inhibitor E64 were added to the resuspended cells. Eventually, the cell suspension was dripped into liquid nitrogen with a 25ml glass pipet and stored at –80°C until further use.

4.2.13 Preparation of Whole-Cell *Tetrahymena* Extracts

For the preparation of high-quality *Tetrahymena* extracts, the desired amount of the frozen cell suspension was weighted-in and bathed with cold RES buffer (see above) supplemented with 5mM DTT (1 ml/mg material). The cells were then gently lysed by thawing in a water bath heated to 42°C. Prior to pelleting the cell debris at 35,000 rpm in a T647.5 rotor (see Materials) for >2 hours at 4°C, 100µM E64 was added to inactive the cysteine proteases set free during lysis. The supernatant was carefully taken off, aliquoted, snap-frozen in liquid nitrogen and stored at –80°C until further use. Note that for quality control, a reporter assay (as shown in Figure 5.4.A) was performed with each freshly-prepared extract. For this purpose, 4µM of an NLS-Sp68-MBP-mCherry fusion were incubated with the extract for 2-4 hours at 30°C to assay protease contamination and analyzed with SDS-PAGE.

4.2.14 Pulldowns for the Identification of Nucleus-specific *Tetrahymena* NTRs

In the present study, three strategies for the identification of nucleus-specific NTRs and NTR•cargo complexes were followed: pulldowns with (i) the TtMacNup98A- and TtMicNup98A-derived GLFG- and IF-repeats, (ii) the putative IBBs of the major MAC- and MIC-specific import adaptors and (iii) the nucleus-specific linker histones. Please note that the repeats were covalently-coupled to Sepharose as described above, whereas the import adaptors and histones were immobilized to IgG-Sepharose via ZZ-tags. For the latter, 60µg protein were incubated for 1 hour at 4°C with IgG-Sepharose, which was pre-eluted with glycine buffer (pH2.2) and equilibrated in 44mM Tris pH7.5, 290mM NaCl and 4.4mM $MgCl_2$.

Irrespective of their coupling, all baits were washed with RES buffer (see above) prior to the addition of the *Tetrahymena* cell extract. Incubation of the extracts (from approx. 20-25 Mio. cells) and the immobilized baits was performed for 1-4 hours at 4°C while gently rotating. Afterwards, the matrices were carefully washed with RES and TRES (50mM Tris pH7.5, 200mM NaCl, 10mM $MgCl_2$, 5mM EDTA) buffer by pipetting up and down with a 1000µl tip. For elution from the covalently-coupled repeats, SDS sample buffer was used. In contrast, bound preys were separated from the ZZ-tagged baits with the help of 1-2M $MgCl_2$. Afterwards, the immobilized baits were released from the IgG-Sepharose by elution with SDS sample buffer lacking DTT. The respective controls of each experiment (see figure legends) were always treated equally.

Finally, all samples were analyzed with SDS-PAGE (using either a custom in-house or the NuPAGE 4-12% Bis-Tris gel system from Invitrogen, Darmstadt, Germany) and stained with colloidal Coomassie (see Materials). For identification of the preys, complete gel lanes or individual bands were cut out and analyzed with mass spectrometry (see below).

4.2.15 Pulldowns for the Identification of additional FG repeats in *Tetrahymena*

For the identification of additional FG repeats in *Tetrahymena*, 130µg of the biotinylated version of the human $Imp\beta_{45\text{-}462}$ mutant described in Section 2.4.8 was bound to Streptavidin-Agarose equilibrated in RES buffer (see above) for 1 at 4°C. Afterwards, ~82.5µg biotin were added to saturate the remaining binding sites of the matrix and the coupling reaction was further incubated for 15 minutes. A likewise passivated Streptavidin-Agarose control was prepared alongside. Next, the matrices were extensively washed with RES buffer to remove all unbound material. *Tetrahymena* extracts (from approx. 20-25 Mio. cells) were then added and incubated with the baits for 2 hours at 4°C while rotating. Prior to elution, all samples were washed with RES and TRES (see above) buffer. The bound material was eluted with hot SDS sample buffer, analyzed with SDS-PAGE (using either a custom in-house or the NuPAGE 4-12% Bis-Tris gel system from Invitrogen, Darmstadt, Germany) and stained with colloidal Coomassie (see Materials). For identification of the preys, complete gel lanes were cut out and analyzed with mass spectrometry (see below).

4.2.16 Pulldowns with the WT and Y-FG mutated Sup35 PrD

For the pulldown experiments with the Sup35 WT and Y-FG PrDs, cytoplasmic HeLa extracts prepared according to Dignam et al., 1983 (kind gift of P. Odenwälder) were used. The protein baits were covalently coupled to Maleimeide-Sepharose as described above. Prior to use, the baits were washed with LS buffer (44mM Tris pH7.5, 290mM NaCl and 4.4mM $MgCl_2$). Eventually, the extracts and immobilized prion domains were incubated for 2 hours at 4°C while gently rotating. Afterwards, unbound material was carefully washed off with LS buffer. The final elution of the bound preys was performed with hot SDS sample buffer. After analysis with SDS-PAGE (using either a custom in-house or the NuPAGE 4-12% Bis-Tris gel system from Invitrogen, Darmstadt, Germany) and staining with colloidal Coomassie (see Materials), complete lanes were cut out and the therein included proteins identified with mass spectrometry (see below).

4.2.17 Mass Spectrometry Analysis of Pulldown Experiments

For protein identification via mass spectrometry, either individual lanes of colloidal Coomassie-stained gels were cut into 23 equally sized slices irrespective of the visible band pattern (complete analysis) or defined single bands were precisely excised (single band analysis). In both cases, in-gel tryptic digestion and peptide extraction were performed as previously described (Shevchenko et al., 1996; Schmidt and Urlaub, 2009). The tryptic peptides from each band or slice were further separated by reverse-phase HPLC on a Reprosil C18 column (Dr. Maisch, Ammerbuch-Entringen, Germany) and directly eluted into an LTQ-Orbitrap XL mass spectrometer (Thermo Fisher Scientific, Dreieich, Germany). MS/MS spectra were recorded and searched against the NCBI non-redundant database with appropriate taxonomy filters (i.e. depending on the individual pulldown experiment analyzed either *Tetrahymena*, Metazoa or *E. coli*; in few cases also without taxonomy filter) with MASCOT v2.2. The MASCOT search results were directly analyzed as the html output files or using Scaffold 3.0 (Proteome Software, Portland, Oregon, USA). The following criteria were applied: peptide identifications were only accepted if their identification probability was greater than 90% as judged by the PeptideProphet algorithm (Keller et al., 2002). Proteins were only considered to be identified when at least two unique peptides have been found and the identification probability exceeded 90% as predicted by the ProteinProphet algorithm (Nesvizhskii et al., 2003). For each accepted protein hit, an individual composite score was calculated by dividing the number of unique spectra obtained for that hit with its molecular weight. Individual experiments were qualitatively compared based on the ratio of this score for a given hit in a pulldown sample and the corresponding control.

Bibliography

Ader, C., Frey, S., Maas, W., Schmidt, H.B., Görlich, D., and Baldus, M. (2010). Amyloid-like interactions within nucleoporin FG hydrogels. Proceedings of the National Academy of Sciences *107*, 6281–6285.

Adessi, C.L., and Soto, C. (2002). Beta-sheet breaker strategy for the treatment of Alzheimer's disease. Drug Dev. Res. *56*, 184–193.

Adl, S.M., et al. (2007). Diversity, nomenclature, and taxonomy of protists. *56*, 684–689.

Akey, C.W. (1995). Structural plasticity of the nuclear pore complex. J Mol Biol *248*, 273–293.

Akey, C.W., and Radermacher, M. (1993). Architecture of the Xenopus Nuclear Pore Complex Revealed by Three-Dimensional Cryo-Electron Microscopy. J Cell Biol *122*, 1–19.

Alber, F., Dokudovskaya, S., Veenhoff, L.M., Zhang, W., Kipper, J., Devos, D., Suprapto, A., Karni-Schmidt, O., Williams, R., Chait, B.T., et al. (2007). The molecular architecture of the nuclear pore complex. Nature *450*, 695–701.

Alberti, S., Halfmann, R., King, O., Kapila, A., and Lindquist, S. (2009). A systematic survey identifies prions and illuminates sequence features of prionogenic proteins. Cell *137*, 146–158.

Alberts, B., Johnson, A., Lewis, J., Raff, M., Roberts, K., and Walter, P. (2002). Molecular Biology of the Cell, Fourth Edition (Garland Science).

Allis, C.D., Allen, R.L., Wiggins, J.C., Chicoine, L.G., and Richman, R. (1984). Proteolytic processing of h1-like histones in chromatin: a physiologically and developmentally regulated event in Tetrahymena micronuclei. J Cell Biol *99*, 1669–1677.

Andrade, M.A., Petosa, C., O'Donoghue, S.I., Müller, C.W., and Bork, P. (2001). Comparison of ARM and HEAT protein repeats. J Mol Biol *309*, 1–18.

Baake, M., Bäuerle, M., Doenecke, D., and Albig, W. (2001). Core histones and linker histones are imported into the nucleus by different pathways. Eur. J. Cell Biol. *80*, 669–677.

Bapteste, E., Charlebois, R.L., MacLeod, D., and Brochier, C. (2005). The two tempos of nuclear pore complex evolution: highly adapting proteins in an ancient frozen structure. Genome Biol *6*, R85.

Bastos, R., Ribas de Pouplana, L., Enarson, M., Bodoor, K., and Burke, B. (1997). Nup84, a novel nucleoporin that is associated with CAN/Nup214 on the cytoplasmic face of the nuclear pore complex. J Cell Biol *137*, 989–1000.

Bayliss, R., Leung, S.W., Baker, R.P., Quimby, B.B., Corbett, A.H., and Stewart, M. (2002a). Structural basis for the interaction between NTF2 and nucleoporin FxFG repeats. Embo J. *21*, 2843–2853.

Bayliss, R., Littlewood, T., and Stewart, M. (2000). Structural basis for the interaction between FxFG nucleoporin repeats and importin-beta in nuclear trafficking. Cell *102*, 99–108.

Bayliss, R., Littlewood, T., Strawn, L.A., Wente, S.R., and Stewart, M. (2002b). GLFG and FxFG nucleoporins bind to overlapping sites on importin-beta. J Biol Chem *277*, 50597–50606.

Bayliss, R., Ribbeck, K., Akin, D., Kent, H.M., Feldherr, C.M., Görlich, D., and Stewart, M. (1999). Interaction between NTF2 and xFxFG-containing nucleoporins is required to mediate nuclear import of RanGDP. J Mol Biol *293*, 579–593.

Beck, M., Förster, F., Ecke, M., Plitzko, J.M., Melchior, F., Gerisch, G., Baumeister, W., and Medalia, O. (2004). Nuclear pore complex structure and dynamics revealed by cryoelectron tomography. Science *306*, 1387–1390.

Beck, M., Lucić, V., Förster, F., Baumeister, W., and Medalia, O. (2007). Snapshots of nuclear pore complexes in action captured by cryo-electron tomography. Nature *449*, 611–615.

Bednenko, J., Cingolani, G., and Gerace, L. (2003). Importin beta contains a COOH-terminal nucleoporin binding region important for nuclear transport. J Cell Biol *162*, 391–401.

Belgareh, N., Snay-Hodge, C., Pasteau, F., Dagher, S., Cole, C.N., and Doye, V. (1998). Functional characterization of a Nup159p-containing nuclear pore subcomplex. Mol Biol Cell *9*, 3475–3492.

Bernad, R., van der Velde, H., Fornerod, M., and Pickersgill, H. (2004). Nup358/RanBP2 attaches to the nuclear pore complex via association with Nup88 and Nup214/CAN and plays a supporting role in CRM1-mediated nuclear protein export. Mol Cell Biol *24*, 2373–2384.

Bischoff, F.R., and Görlich, D. (1997). RanBP1 is crucial for the release of RanGTP from importin beta-related nuclear transport factors. FEBS Lett *419*, 249–254.

Bischoff, F.R., and Ponstingl, H. (1991a). Catalysis of guanine nucleotide exchange on Ran by the mitotic regulator RCC1. Nature *354*, 80–82.

Bischoff, F.R., and Ponstingl, H. (1991b). Mitotic regulator protein RCC1 is complexed with a nuclear ras-related polypeptide. Proc Natl Acad Sci USA *88*, 10830–10834.

Bischoff, F.R., Klebe, C., Kretschmer, J., Wittinghofer, A., and Ponstingl, H. (1994). RanGAP1 induces GTPase activity of nuclear Ras-related Ran. Proc Natl Acad Sci USA *91*, 2587–2591.

Bonner, W.M. (1975). Protein migration into nuclei. I. Frog oocyte nuclei in vivo accumulate microinjected histones, allow entry to small proteins, and exclude large proteins. J Cell Biol *64*, 421–430.

Bright, J.N., Woolf, T.B., and Hoh, J.H. (2001). Predicting properties of intrinsically unstructured proteins. Prog. Biophys. Mol. Biol. *76*, 131–173.

Brock, J.E., Paz, R.L., Cottle, P., and Janssen, G.R. (2007). Naturally occurring adenines within mRNA coding sequences affect ribosome binding and expression in Escherichia coli. J Bacteriol *189*, 501–510.

Brohawn, S.G., and Schwartz, T.U. (2009). Molecular architecture of the Nup84-Nup145C-Sec13 edge element in the nuclear pore complex lattice. Nat Struct Mol Biol *16*, 1173–1177.

Brohawn, S.G., Leksa, N.C., Spear, E.D., Rajashankar, K.R., and Schwartz, T.U. (2008). Structural evidence for common ancestry of the nuclear pore complex and vesicle coats. Science *322*, 1369–1373.

Brohawn, S.G., Partridge, J.R., Whittle, J.R.R., and Schwartz, T.U. (2009). The nuclear pore complex has entered the atomic age. Structure *17*, 1156–1168.

Bullock, T.L., Clarkson, W.D., Kent, H.M., and Stewart, M. (1996). The 1.6 angstroms resolution crystal structure of nuclear transport factor 2 (NTF2). J Mol Biol *260*, 422–431.

Buss, F., and Stewart, M. (1995). Macromolecular interactions in the nucleoporin p62 complex of rat nuclear pores: binding of nucleoporin p54 to the rod domain of p62. J Cell Biol *128*, 251–261.

Byrd, D.A., Sweet, D.J., Panté, N., Konstantinov, K.N., Guan, T., Saphire, A.C., Mitchell, P.J., Cooper, C.S., Aebi, U., and Gerace, L. (1994). Tpr, a large coiled coil protein whose amino terminus is involved in activation of oncogenic kinases, is localized to the cytoplasmic surface of the nuclear pore complex. J Cell Biol *127*, 1515–1526.

Cassidy-Hanley, D.M. (2012). Tetrahymena in the laboratory: strain resources, methods for culture, maintenance, and storage. Methods Cell Biol *109*, 237–276.

Cavalier-Smith, T. (2001). Obcells as proto-organisms: membrane heredity, lithophosphorylation, and the origins of the genetic code, the first cells, and photosynthesis. Journal of Molecular Evolution *53*, 555–595.

Cavalier-Smith, T. (2010a). Kingdoms Protozoa and Chromista and the eozoan root of the eukaryotic tree. Biol Lett *6*, 342–345.

Cavalier-Smith, T. (2010b). Origin of the cell nucleus, mitosis and sex: roles of intracellular coevolution. Biology Direct *5*, 7.

Cervantes, M.D., Xi, X., Vermaak, D., Yao, M.-C., and Malik, H.S. (2006). The CNA1 histone of the ciliate Tetrahymena thermophila is essential for chromosome segregation in the germline micronucleus. Mol Biol Cell *17*, 485–497.

Chalker, D.L. (2008). Dynamic nuclear reorganization during genome remodeling of Tetrahymena. Biochim Biophys Acta *1783*, 2130–2136.

Cheng, Z., Saito, K., Pisarev, A.V., Wada, M., Pisareva, V.P., Pestova, T.V., Gajda, M., Round, A., Kong, C., Lim, M., et al. (2009). Structural insights into eRF3 and stop codon recognition by eRF1. Genes Dev. *23*, 1106–1118.

Chenna, R., Sugawara, H., Koike, T., Lopez, R., Gibson, T.J., Higgins, D.G., and Thompson, J.D. (2003). Multiple sequence alignment with the Clustal series of programs. Nucleic Acids Res. *31*, 3497–3500.

Chernoff, Y.O., Lindquist, S.L., Ono, B., Inge-Vechtomov, S.G., and Liebman, S.W. (1995). Role of the chaperone protein Hsp104 in propagation of the yeast prion-like factor [psi+]. Science *268*, 880–884.

Chien, P., Weissman, J.S., and DePace, A.H. (2004). Emerging principles of conformation-based prion inheritance. Annu Rev Biochem *73*, 617–656.

Chiti, F., and Dobson, C.M. (2006). Protein misfolding, functional amyloid, and human disease. Annu Rev Biochem *75*, 333–366.

Cingolani, G., Petosa, C., Weis, K., and Müller, C.W. (1999). Structure of importin-beta bound to the IBB domain of importin-alpha. Nature *399*, 221–229.

Cohen, J. (1988). Statistical power analysis for the behavioral sciences (Hillsdale, New Jersey: Lawrence Erlbaum Associates Inc.).

Cole, C., Barber, J.D., and Barton, G.J. (2008). The Jpred 3 secondary structure prediction server. Nucleic Acids Res. *36*, W197–W201.

Colwell, L.J., Brenner, M.P., and Ribbeck, K. (2010). Charge as a selection criterion for translocation through the nuclear pore complex. PLoS Comput Biol *6*, e1000747.

Cook, A., Bono, F., Jinek, M., and Conti, E. (2007). Structural biology of nucleocytoplasmic transport. Annu Rev Biochem *76*, 647–671.

Cronshaw, J.M., Krutchinsky, A.N., Zhang, W., Chait, B.T., and Matunis, M.J. (2002). Proteomic analysis of the mammalian nuclear pore complex. J Cell Biol *158*, 915–927.

Dangelo, M., and Hetzer, M. (2008). Structure, dynamics and function of nuclear pore complexes. Trends in Cell Biology *18*, 456–466.

Davis, L.I., and Blobel, G. (1987). Nuclear pore complex contains a family of glycoproteins that includes p62: glycosylation through a previously unidentified cellular pathway. Proc Natl Acad Sci USA *84*, 7552–7556.

Davis, L.I., and Fink, G.R. (1990). The NUP1 gene encodes an essential component of the yeast nuclear pore complex. Cell *61*, 965–978.

Deane, R., Schäfer, W., Zimmermann, H.P., Mueller, L., Görlich, D., Prehn, S., Ponstingl, H., and Bischoff, F.R. (1997). Ran-binding protein 5 (RanBP5) is related to the nuclear transport factor importin-beta but interacts differently with RanBP1. Mol Cell Biol *17*, 5087–5096.

Degrasse, J., Dubois, K., Devos, D., Siegel, T., Sali, A., Field, M., Rout, M., and Chait, B. (2009). Evidence for a Shared Nuclear Pore Complex Architecture That Is Conserved from the Last Common Eukaryotic Ancestor. Molecular & Cellular Proteomics *8*, 2119.

Denning, D.P., and Rexach, M.F. (2007). Rapid evolution exposes the boundaries of domain structure and function in natively unfolded FG nucleoporins. Mol Cell Proteomics *6*, 272–282.

Denning, D.P., Patel, S.S., Uversky, V., Fink, A.L., and Rexach, M. (2003). Disorder in the nuclear pore complex: the FG repeat regions of nucleoporins are natively unfolded. Proc Natl Acad Sci USA *100*, 2450–2455.

DePace, A.H., Santoso, A., Hillner, P., and Weissman, J.S. (1998). A critical role for amino-terminal glutamine/asparagine repeats in the formation and propagation of a yeast prion. Cell *93*, 1241–1252.

Devos, D., Dokudovskaya, S., Alber, F., Williams, R., Chait, B.T., Sali, A., and Rout, M.P. (2004). Components of coated vesicles and nuclear pore complexes share a common molecular architecture. PLoS Biol *2*, e380.

Dignam, J.D., Lebovitz, R.M., and Roeder, R.G. (1983). Accurate transcription initiation by RNA polymerase II in a soluble extract from isolated mammalian nuclei. Nucleic Acids Res. *11*, 1475–1489.

Dokyun Na, S.L.D.L. (2010). Mathematical modeling of translation initiation for the estimation of its efficiency to computationally design mRNA sequences with desired expression levels in prokaryotes. BMC Systems Biology *4*, 71.

Dölker, N., Zachariae, U., and Grubmüller, H. (2010). Hydrophilic linkers and polar contacts affect aggregation of FG repeat peptides. Biophys J *98*, 2653–2661.

Eisele, N.B., Frey, S., Piehler, J., Görlich, D., and Richter, R.P. (2010). Ultrathin nucleoporin phenylalanine-glycine repeat films and their interaction with nuclear transport receptors. EMBO Rep. *11*, 366–372.

Eisen, J.A., Coyne, R.S., Wu, M., Wu, D., Thiagarajan, M., Wortman, J.R., Badger, J.H., Ren, Q., Amedeo, P., Jones, K.M., et al. (2006). Macronuclear genome sequence of the ciliate Tetrahymena thermophila, a model eukaryote. PLoS Biol *4*, e286.

Fauchere, J.L., and Pliska, VE (1983). Hydrophobic parameters of amino-acid side-chains from the partitioning of N-acetyl-amino-acid amide. Eur J Med Chem *18*, 369–375.

Fayyad, U., Piatetsky-Shapiro, G., and Smyth, P. (1996). From data mining to knowledge discovery in databases. AI Magazine *17*, 37.

Feldherr, C.M., Kallenbach, E., and Schultz, N. (1984). Movement of a karyophilic protein through the nuclear pores of oocytes. J Cell Biol *99*, 2216–2222.

Field, M.C., Sali, A., and Rout, M.P. (2011). Evolution: On a bender--BARs, ESCRTs, COPs, and finally getting your coat. J Cell Biol *193*, 963–972.

Finlay, D.R., Meier, E., Bradley, P., Horecka, J., and Forbes, D.J. (1991). A complex of nuclear pore proteins required for pore function. J Cell Biol *114*, 169–183.

Fontoura, B.M., Blobel, G., and Matunis, M.J. (1999). A conserved biogenesis pathway for nucleoporins: proteolytic processing of a 186-kilodalton precursor generates Nup98 and the novel nucleoporin, Nup96. J Cell Biol *144*, 1097–1112.

Fornerod, M., Ohno, M., Yoshida, M., and Mattaj, I.W. (1997a). CRM1 is an export receptor for leucine-rich nuclear export signals. Cell *90*, 1051–1060.

Fornerod, M., van Deursen, J., van Baal, S., Reynolds, A., Davis, D., Murti, K.G., Fransen, J., and Grosveld, G. (1997b). The human homologue of yeast CRM1 is in a dynamic subcomplex with CAN/Nup214 and a novel nuclear pore component Nup88. Embo J. *16*, 807–816.

Frenkiel-Krispin, D., Maco, B., Aebi, U., and Medalia, O. (2010). Structural analysis of a metazoan nuclear pore complex reveals a fused concentric ring architecture. J Mol Biol *395*, 578–586.

Frey, S., and Görlich, D. (2007). A saturated FG-repeat hydrogel can reproduce the permeability properties of nuclear pore complexes. Cell *130*, 512–523.

Frey, S., and Görlich, D. (2009). FG/FxFG as well as GLFG repeats form a selective permeability barrier with self-healing properties. Embo J. *28*, 2554–2567.

Frey, S., Richter, R.P., and Görlich, D. (2006). FG-rich repeats of nuclear pore proteins form a three-dimensional meshwork with hydrogel-like properties. Science *314*, 815–817.

Fried, H., and Kutay, U. (2003). Nucleocytoplasmic transport: taking an inventory. Cellular and Molecular Life Sciences (CMLS) *60*, 1659–1688.

Fuerst, J.A. (2005). Intracellular compartmentation in planctomycetes. Annu Rev Microbiol *59*, 299–328.

Gall, J.G. (1967). Octagonal nuclear pores. J Cell Biol *32*, 391–399.

Gerace, L., Ottaviano, Y., and Kondor-Koch, C. (1982). Identification of a major polypeptide of the nuclear pore complex. J Cell Biol *95*, 826–837.

Gilbert, W. (1978). Why genes in pieces? Nature *271*, 501.

Godiska, R., Mead, D., Dhodda, V., Wu, C., Hochstein, R., Karsi, A., Usdin, K., Entezam, A., and Ravin, N. (2010). Linear plasmid vector for cloning of repetitive or unstable sequences in Escherichia coli. Nucleic Acids Res. *38*, e88.

Goldberg, M.W., and Allen, T.D. (1992). High resolution scanning electron microscopy of the nuclear envelope: demonstration of a new, regular, fibrous lattice attached to the baskets of the nucleoplasmic face of the nuclear pores. J Cell Biol *119*, 1429–1440.

Gorovsky, M.A., Glover, C., Johmann, C.A., Keevert, J.B., Mathis, D.J., and Samuelson, M. (1978). Histones and chromatin structure in Tetrahymena macro- and micronuclei. Cold Spring Harb. Symp. Quant. Biol. *42 Pt 1*, 493–503.

Gorovsky, M.A., Keevert, J.B., and Pleger, G.L. (1974). Histone F1 of Tetrahymena macronuclei: unique electrophoretic properties and phosphorylation of F1 in an amitotic nucleus. J Cell Biol *61*, 134–145.

Görlich, D. (1997). A Novel Class of RanGTP Binding Proteins. J Cell Biol *138*, 65–80.

Görlich, D., and Kutay, U. (1999). Transport between the cell nucleus and the cytoplasm. Annu. Rev. Cell Dev. Biol. *15*, 607–660.

Görlich, D., Dabrowski, M., Bischoff, F.R., Kutay, U., Bork, P., Hartmann, E., Prehn, S., and Izaurralde, E. (1997). A novel class of RanGTP binding proteins. J Cell Biol *138*, 65–80.

Görlich, D., Henklein, P., Laskey, R.A., and Hartmann, E. (1996a). A 41 amino acid motif in importin-alpha confers binding to importin-beta and hence transit into the nucleus. Embo J. *15*, 1810–1817.

Görlich, D., Kostka, S., Kraft, R., Dingwall, C., Laskey, R.A., Hartmann, E., and Prehn, S. (1995a). Two different subunits of importin cooperate to recognize nuclear localization signals and bind them to the nuclear envelope. Curr Biol *5*, 383–392.

Görlich, D., Panté, N., Kutay, U., Aebi, U., and Bischoff, F.R. (1996b). Identification of different roles for RanGDP and RanGTP in nuclear protein import. Embo J. *15*, 5584–5594.

Görlich, D., Prehn, S., Laskey, R.A., and Hartmann, E. (1994). Isolation of a protein that is essential for the first step of nuclear protein import. Cell *79*, 767–778.

Görlich, D., Seewald, M.J., and Ribbeck, K. (2003). Characterization of Ran-driven cargo transport and the RanGTPase system by kinetic measurements and computer simulation. Embo J. *22*, 1088–1100.

Görlich, D., Vogel, F., Mills, A.D., Hartmann, E., and Laskey, R.A. (1995b). Distinct functions for the two importin subunits in nuclear protein import. Nature *377*, 246–248.

Grandi, P., Dang, T., Pané, N., Shevchenko, A., Mann, M., Forbes, D., and Hurt, E. (1997). Nup93, a vertebrate homologue of yeast Nic96p, forms a complex with a novel 205-kDa protein and is required for correct nuclear pore assembly. Mol Biol Cell *8*, 2017–2038.

Grandi, P., Schlaich, N., Tekotte, H., and Hurt, E.C. (1995). Functional interaction of Nic96p with a core nucleoporin complex consisting of Nsp1p, Nup49p and a novel protein Nup57p. Embo J. *14*, 76–87.

Grossman, E., Medalia, O., and Zwerger, M. (2012). Functional architecture of the nuclear pore complex. Annu Rev Biophys *41*, 557–584.

Güttler, T., and Görlich, D. (2011). Ran-dependent nuclear export mediators: a structural perspective. Embo J. *30*, 3457–3474.

Güttler, T., Madl, T., Neumann, P., Deichsel, D., Corsini, L., Monecke, T., Ficner, R., Sattler, M., and Görlich, D. (2010). NES consensus redefined by structures of PKI-type and Rev-type nuclear export signals bound to CRM1. Nat Struct Mol Biol *17*, 1367–1376.

Halfmann, R., Alberti, S., and Lindquist, S. (2010). Prions, protein homeostasis, and phenotypic diversity. Trends in Cell Biology *20*, 125–133.

Halfmann, R., Alberti, S., Krishnan, R., Lyle, N., O'Donnell, C.W., King, O.D., Berger, B., Pappu, R.V., and Lindquist, S. (2011). Opposing effects of glutamine and asparagine govern prion formation by intrinsically disordered proteins. Mol Cell *43*, 72–84.

Halfmann, R., Jarosz, D.F., Jones, S.K., Chang, A., Lancaster, A.K., and Lindquist, S. (2012a). Prions are a common mechanism for phenotypic inheritance in wild yeasts. Nature *482*, 363–368.

Halfmann, R., Wright, J., Alberti, S., Lindquist, S., and Rexach, M. (2012b). Prion formation by a yeast GLFG nucleoporin. Prion *6*.

Hallberg, E., Wozniak, R.W., and Blobel, G. (1993). An integral membrane protein of the pore membrane domain of the nuclear envelope contains a nucleoporin-like region. J Cell Biol *122*, 513–521.

Harding, C.V., and Feldherr, C. (1958). Semipermeability of the nuclear membrane. Nature *182*, 676–677.

Harrison, P.M., and Gerstein, M. (2003). A method to assess compositional bias in biological sequences and its application to prion-like glutamine/asparagine-rich domains in eukaryotic proteomes. Genome Biol *4*, R40.

Hawryluk-Gara, L.A., Shibuya, E.K., and Wozniak, R.W. (2005). Vertebrate Nup53 interacts with the nuclear lamina and is required for the assembly of a Nup93-containing complex. Mol Biol Cell *16*, 2382–2394.

Hellenbroich, D., Valley, U., Ryll, T., Wagner, R., Tekkanat, N., Kessler, W., Ross, A., and Deckwer, W.D. (1999). Cultivation of Tetrahymena thermophila in a 1.5-m3 airlift bioreactor. Appl Microbiol Biotechnol *51*, 447–455.

Hinshaw, J.E., Carragher, B.O., and Milligan, R.A. (1992). Architecture and design of the nuclear pore complex. Cell *69*, 1133–1141.

Hodel, A.E., Hodel, M.R., Griffis, E.R., Hennig, K.A., Ratner, G.A., Xu, S., and Powers, M.A. (2002). The three-dimensional structure of the autoproteolytic, nuclear pore-targeting domain of the human nucleoporin Nup98. Mol Cell *10*, 347–358.

Hou, F., Sun, L., Zheng, H., Skaug, B., Jiang, Q.-X., and Chen, Z.J. (2011). MAVS forms functional prion-like aggregates to activate and propagate antiviral innate immune response. Cell *146*, 448–461.

Hu, T., Guan, T., and Gerace, L. (1996). Molecular and functional characterization of the p62 complex, an assembly of nuclear pore complex glycoproteins. J Cell Biol *134*, 589–601.

Hurtado-Guerrero, R., Dorfmueller, H.C., and van Aalten, D.M. (2008). Molecular mechanisms of O-GlcNAcylation. Current Opinion in Structural Biology *18*, 551–557.

Hülsmann, B.B., Labokha, A.A., and Görlich, D. (2012). The permeability of reconstituted nuclear pores provides direct evidence for the selective phase model. Cell *150*, 738–751.

Imamoto, N., Tachibana, T., Matsubae, M., and Yoneda, Y. (1995). A karyophilic protein forms a stable complex with cytoplasmic components prior to nuclear pore binding. J Biol Chem *270*, 8559–8565.

Iovine, M.K., Watkins, J.L., and Wente, S.R. (1995). The GLFG repetitive region of the nucleoporin Nup116p interacts with Kap95p, an essential yeast nuclear import factor. J Cell Biol *131*, 1699–1713.

Isgro, T.A., and Schulten, K. (2005). Binding dynamics of isolated nucleoporin repeat regions to importin-beta. Structure *13*, 1869–1879.

Iwamoto, M., Mori, C., Kojidani, T., Bunai, F., Hori, T., Fukagawa, T., Hiraoka, Y., and Haraguchi, T. (2009). Two distinct repeat sequences of Nup98 nucleoporins characterize dual nuclei in the binucleated ciliate tetrahymena. Curr Biol *19*, 843–847.

Izaurralde, E., Kutay, U., Kobbe, von, C., Mattaj, I.W., and Görlich, D. (1997). The asymmetric distribution of the constituents of the Ran system is essential for transport into and out of the nucleus. Embo J. *16*, 6535–6547.

Jaenicke, R., and Seckler, R. (1997). Protein misassembly in vitro. Adv. Protein Chem. *50*, 1–59.

Jarnik, M., and Aebi, U. (1991). Toward a more complete 3-D structure of the nuclear pore complex. J Struct Biol *107*, 291–308.

Jäkel, S., and Görlich, D. (1998). Importin β, transportin, RanBP5 and RanBP7 mediate nuclear import of ribosomal proteins in mammalian cells. Embo J. *17*, 4491–4502.

Johmann, C.A., and Gorovsky, M.A. (1976). Purification and characterization of the histones associated with the macronucleus of Tetrahymena. Biochemistry *15*, 1249–1256.

Jovanovic-Talisman, T., Tetenbaum-Novatt, J., McKenney, A., Zilman, A., Peters, R., Rout, M., and Chait, B. (2008). Artificial nanopores that mimic the transport selectivity of the nuclear pore complex. Nature.

Kalderon, D., Roberts, B.L., Richardson, W.D., and Smith, A.E. (1984). A short amino acid sequence able to specify nuclear location. Cell *39*, 499–509.

Karrer, K.M. (2012). Nuclear Dualism (Elsevier Inc.).

Kaufman, L. (2005). Finding Groups in Data: An Introduction to Cluster Analysis (Wiley-Interscience).

Keeling, P.J., Burger, G., Durnford, D.G., Lang, B.F., Lee, R.W., Pearlman, R.E., Roger, A.J., and Gray, M.W. (2005). The tree of eukaryotes. Trends Ecol Evol (Amst) *20*, 670–676.

Keller, A., Nesvizhskii, A.I., Kolker, E., and Aebersold, R. (2002). Empirical statistical model to estimate the accuracy of peptide identifications made by MS/MS and database search. Anal. Chem. *74*, 5383–5392.

Kiseleva, E., Allen, T.D., Rutherford, S., Bucci, M., Wente, S.R., and Goldberg, M.W. (2004). Yeast nuclear pore complexes have a cytoplasmic ring and internal filaments. J Struct Biol *145*, 272–288.

Kiy, T., and Tiedtke, A. (1992). Continuous high-cell-density fermentation of the ciliated protozoon Tetrahymena in a perfused bioreactor. Appl Microbiol Biotechnol *38*, 141–146.

Klebe, C., Bischoff, F.R., Ponstingl, H., and Wittinghofer, A. (1995). Interaction of the nuclear GTP-binding protein Ran with its regulatory proteins RCC1 and RanGAP1. Biochemistry *34*, 639–647.

Kobe, B. (1999). Autoinhibition by an internal nuclear localization signal revealed by the crystal structure of mammalian importin alpha. Nat Struct Biol *6*, 388–397.

Kong, C., Ito, K., Walsh, M.A., Wada, M., Liu, Y., Kumar, S., Barford, D., Nakamura, Y., and Song, H. (2004). Crystal structure and functional analysis of the eukaryotic class II release factor eRF3 from S. pombe. Mol Cell *14*, 233–245.

Kose, S., Imamoto, N., Tachibana, T., Shimamoto, T., and Yoneda, Y. (1997). Ran-unassisted nuclear migration of a 97-kD component of nuclear pore-targeting complex. J Cell Biol *139*, 841–849.

Kowalczyk, S.W., Kapinos, L., Blosser, T.R., Magalhães, T., van Nies, P., Lim, R.Y.H., and Dekker, C. (2011). Single-molecule transport across an individual biomimetic nuclear pore complex. Nature Nanotechnology 1–6.

Kraemer, D., Wozniak, R.W., Blobel, G., and Radu, A. (1994). The human CAN protein, a putative oncogene product associated with myeloid leukemogenesis, is a nuclear pore complex protein that faces the cytoplasm. Proc Natl Acad Sci USA *91*, 1519–1523.

Kramer, A., Liashkovich, I., Ludwig, Y., and Shahin, V. (2008). Atomic force microscopy visualises a hydrophobic meshwork in the central channel of the nuclear pore. Pflugers Arch. *456*, 155–162.

Krull, S., Dörries, J., Boysen, B., Reidenbach, S., Magnius, L., Norder, H., Thyberg, J., and Cordes, V.C. (2010). Protein Tpr is required for establishing nuclear pore-associated zones of heterochromatin exclusion. Embo J. *29*, 1659–1673.

Krull, S., Thyberg, J., rkroth, B.B., Rackwitz, H.-R., and Cordes, V.C. (2004). Nucleoporins as components of the nuclear pore complex core structure and Tpr as the architectural element of the nuclear basket. Mol Biol Cell *15*, 4261–4277.

Kruskal, W.H., and Wallis, W.A. (1952). Use of Ranks in One-Criterion Variance Analysis. Journal of the American Statistical Association *47*, 583–621.

Kryndushkin, D.S., Alexandrov, I.M., Ter-Avanesyan, M.D., and Kushnirov, V.V. (2003). Yeast [PSI+] prion aggregates are formed by small Sup35 polymers fragmented by Hsp104. J Biol Chem *278*, 49636–49643.

Kudla, G., Murray, A.W., Tollervey, D., and Plotkin, J.B. (2009). Coding-Sequence Determinants of Gene Expression in Escherichia coli. Science *324*, 255–258.

Kushner, A.M., and Guan, Z. (2011). Modular design in natural and biomimetic soft materials. Angew Chem Int Ed Engl *50*, 9026–9057.

Kutay, U., Bischoff, F.R., Kostka, S., Kraft, R., and Görlich, D. (1997a). Export of importin alpha from the nucleus is mediated by a specific nuclear transport factor. Cell *90*, 1061–1071.

Kutay, U., Izaurralde, E., Bischoff, F.R., Mattaj, I.W., and Görlich, D. (1997b). Dominant-negative mutants of importin-beta block multiple pathways of import and export through the nuclear pore complex. Embo J. *16*, 1153–1163.

Kyle, R.A. (2001). Amyloidosis: a convoluted story. Br. J. Haematol. *114*, 529–538.

Laemmli, U.K. (1970). Cleavage of structural proteins during the assembly of the head of bacteriophage T4. Nature *227*, 680–685.

Lancaster, A.K., Bardill, J.P., True, H.L., and Masel, J. (2010). The spontaneous appearance rate of the yeast prion [PSI+] and its implications for the evolution of the evolvability properties of the [PSI+] system. Genetics *184*, 393–400.

Lane, C.E., and Archibald, J.M. (2008). The eukaryotic tree of life: endosymbiosis takes its TOL. Trends Ecol Evol (Amst) *23*, 268–275.

Laurell, E., Beck, K., Krupina, K., Theerthagiri, G., Bodenmiller, B., Horvath, P., Aebersold, R., Antonin, W., and Kutay, U. (2011). Phosphorylation of Nup98 by multiple kinases is crucial for NPC disassembly during mitotic entry. Cell *144*, 539–550.

Leksa, N.C., Brohawn, S.G., and Schwartz, T.U. (2009). The structure of the scaffold nucleoporin Nup120 reveals a new and unexpected domain architecture. Structure *17*, 1082–1091.

Li, L., and Lindquist, S. (2000). Creating a protein-based element of inheritance. Science *287*, 661–664.

Liang, H., Xu, J., Zhao, D., Tian, H., Yang, X., Liang, A., and Wang, W. (2012). Subcellular localization and role of Ran1 in Tetrahymena thermophila amitotic macronucleus. FEBS Journal *279*, 2520–2533.

Lim, R.Y.H., and Deng, J. (2009). Interaction forces and reversible collapse of a polymer brush-gated nanopore. ACS Nano *3*, 2911–2918.

Lim, R.Y.H., Fahrenkrog, B., Köser, J., Schwarz-Herion, K., Deng, J., and Aebi, U. (2007). Nanomechanical basis of selective gating by the nuclear pore complex. Science *318*, 640–643.

Lowe, A.R., Siegel, J.J., Kalab, P., Siu, M., Weis, K., and Liphardt, J.T. (2010). Selectivity mechanism of the nuclear pore complex characterized by single cargo tracking. Nature *467*, 600–603.

López de la Paz, M., and Serrano, L. (2004). Sequence determinants of amyloid fibril formation. Proc Natl Acad Sci USA *101*, 87–92.

Lutzmann, M., Kunze, R., Buerer, A., Aebi, U., and Hurt, E. (2002). Modular self-assembly of a Y-shaped multiprotein complex from seven nucleoporins. Embo J. *21*, 387–397.

Lührs, T., Ritter, C., Adrian, M., Riek-Loher, D., Bohrmann, B., Döbeli, H., Schubert, D., and Riek, R. (2005). 3D structure of Alzheimer's amyloid-beta(1-42) fibrils. Proc Natl Acad Sci USA *102*, 17342–17347.

Maimon, T., Elad, N., Dahan, I., and Medalia, O. (2012). The human nuclear pore complex as revealed by cryo-electron tomography. Structure *20*, 998–1006.

Majumdar, A., Cesario, W.C., White-Grindley, E., Jiang, H., Ren, F., Khan, M.R., Li, L., Choi, E.M.-L., Kannan, K., Guo, F., et al. (2012). Critical role of amyloid-like oligomers of Drosophila Orb2 in the persistence of memory. Cell *148*, 515–529.

Malone, C.D., Falkowska, K.A., Li, A.Y., Galanti, S.E., Kanuru, R.C., LaMont, E.G., Mazzarella, K.C., Micev, A.J., Osman, M.M., Piotrowski, N.K., et al. (2008). Nucleus-specific importin alpha proteins and nucleoporins regulate protein import and nuclear division in the binucleate Tetrahymena thermophila. Eukaryotic Cell *7*, 1487–1499.

Mans, B.J., Anantharaman, V., Aravind, L., and Koonin, E.V. (2004). Comparative genomics, evolution and origins of the nuclear envelope and nuclear pore complex. Cell Cycle *3*, 1612–1637.

Mansfeld, J., Güttinger, S., Hawryluk-Gara, L.A., Pante, N., Mall, M., Galy, V., Haselmann, U., Mühlhäusser, P., Wozniak, R.W., Mattaj, I.W., et al. (2006). The conserved transmembrane nucleoporin NDC1 is required for nuclear pore complex assembly in vertebrate cells. Mol Cell *22*, 93–103.

Mason, D.A., Stage, D.E., and Goldfarb, D.S. (2009). Evolution of the metazoan-specific importin alpha gene family. Journal of Molecular Evolution *68*, 351–365.

Miao, L., and Schulten, K. (2009). Transport-related structures and processes of the nuclear pore complex studied through molecular dynamics. Structure *17*, 449–459.

Miao, L., and Schulten, K. (2010). Probing a Structural Model of the Nuclear Pore Complex Channel through Molecular Dynamics. Biophys J *98*, 1658–1667.

Michelitsch, M.D., and Weissman, J.S. (2000). A census of glutamine/asparagine-rich regions: implications for their conserved function and the prediction of novel prions. Proc Natl Acad Sci USA *97*, 11910–11915.

Miller, B.R., Powers, M., Park, M., Fischer, W., and Forbes, D.J. (2000). Identification of a new vertebrate nucleoporin, Nup188, with the use of a novel organelle trap assay. Mol Biol Cell *11*, 3381–3396.

Milles, S., and Lemke, E.A. (2011). Single Molecule Study of the Intrinsically Disordered FG-Repeat Nucleoporin 153. Biophys J *101*, 1710–1719.

Milner, S.T. (1991). Polymer brushes. Science *251*, 905–914.

Mohr, D., Frey, S., Fischer, T., Güttler, T., and Görlich, D. (2009). Characterisation of the passive permeability barrier of nuclear pore complexes. Embo J. *28*, 2541–2553.

Moore, M.S., and Blobel, G. (1992). The two steps of nuclear import, targeting to the nuclear envelope and translocation through the nuclear pore, require different cytosolic factors. Cell *69*, 939–950.

Morrison, J., Yang, J.-C., Stewart, M., and Neuhaus, D. (2003). Solution NMR study of the interaction between NTF2 and nucleoporin FxFG repeats. J Mol Biol *333*, 587–603.

Mosammaparast, N., Guo, Y., Shabanowitz, J., Hunt, D.F., and Pemberton, L.F. (2002). Pathways mediating the nuclear import of histones H3 and H4 in yeast. J Biol Chem *277*, 862–868.

Mosammaparast, N., Jackson, K.R., Guo, Y., Brame, C.J., Shabanowitz, J., Hunt, D.F., and Pemberton, L.F. (2001). Nuclear import of histone H2A and H2B is mediated by a network of karyopherins. J Cell Biol *153*, 251–262.

Moussavi-Baygi, R., Jamali, Y., Karimi, R., and Mofrad, M.R.K. (2011). Brownian dynamics simulation of nucleocytoplasmic transport: a coarse-grained model for the functional state of the nuclear pore complex. PLoS Comput Biol *7*, e1002049.

Mullis, K., Faloona, F., Scharf, S., Saiki, R., Horn, G., and Erlich, H. (1986). Specific enzymatic amplification of DNA in vitro: the polymerase chain reaction. Cold Spring Harb. Symp. Quant. Biol. *51 Pt 1*, 263–273.

Mühlhäusser, P., Müller, E.C., Otto, A., and Kutay, U. (2001). Multiple pathways contribute to nuclear import of core histones. EMBO Rep. *2*, 690–696.

Nachury, M.V., and Weis, K. (1999). The direction of transport through the nuclear pore can be inverted. Proc Natl Acad Sci USA *96*, 9622–9627.

Nagata, K., Takemasa, T., Alam, S., Hattori, T., Watanabe, Y., and Nozawa, Y. (1994). Cloning of cDNAs encoding a cell-cycle-regulatory GTP-binding low-M(r) (GBLM) protein, Ran/TC4, from micronucleated Tetrahymena thermophila and amicronucleated Tetrahymena pyriformis. Gene *144*, 123–125.

Nehrbass, U., Kern, H., Mutvei, A., Horstmann, H., Marshallsay, B., and Hurt, E.C. (1990). NSP1: a yeast nuclear envelope protein localized at the nuclear pores exerts its essential function by its carboxy-terminal domain. Cell *61*, 979–989.

Nelson, R., Sawaya, M.R., Balbirnie, M., Madsen, A.Ø., Riekel, C., Grothe, R., and Eisenberg, D. (2005). Structure of the cross-beta spine of amyloid-like fibrils. Nature *435*, 773–778.

Nesvizhskii, A.I., Keller, A., Kolker, E., and Aebersold, R. (2003). A statistical model for identifying proteins by tandem mass spectrometry. Anal. Chem. *75*, 4646–4658.

Neumann, N., (null), and Poole, A.M. (2010). Comparative genomic evidence for a complete nuclear pore complex in the last eukaryotic common ancestor. PLoS ONE *5*, e13241.

Nguyen Ba, A.N., Pogoutse, A., Provart, N., and Moses, A.M. (2009). NLStradamus: a simple Hidden Markov Model for nuclear localization signal prediction. BMC Bioinformatics *10*, 202.

Onischenko, E., Stanton, L.H., Madrid, A.S., Kieselbach, T., and Weis, K. (2009). Role of the Ndc1 interaction network in yeast nuclear pore complex assembly and maintenance. J Cell Biol *185*, 475–491.

Orias, E., Cervantes, M.D., and Hamilton, E.P. (2011). Tetrahymena thermophila, a unicellular eukaryote with separate germline and somatic genomes. Res Microbiol 1–10.

Osherovich, L.Z., Cox, B.S., Tuite, M.F., and Weissman, J.S. (2004). Dissection and design of yeast prions. PLoS Biol *2*, E86.

Partridge, J.R., and Schwartz, T.U. (2009). Crystallographic and biochemical analysis of the Ran-binding zinc finger domain. J Mol Biol *391*, 375–389.

Paschal, B.M., and Gerace, L. (1995). Identification of NTF2, a cytosolic factor for nuclear import that interacts with nuclear pore complex protein p62. J Cell Biol *129*, 925–937.

Patel, S., Belmont, B., Sante, J., and Rexach, M. (2007). Natively Unfolded Nucleoporins Gate Protein Diffusion across the Nuclear Pore Complex. Cell *129*, 83–96.

Patino, M.M., Liu, J.J., Glover, J.R., and Lindquist, S. (1996). Support for the prion hypothesis for inheritance of a phenotypic trait in yeast. Science *273*, 622–626.

Permanne, B., Adessi, C., Fraga, S., Frossard, M.J., Saborio, G.P., and Soto, C. (2002). Are beta-sheet breaker peptides dissolving the therapeutic problem of Alzheimer's disease? J. Neural Transm. Suppl. 293–301.

Perutz, M.F., Pope, B.J., Owen, D., Wanker, E.E., and Scherzinger, E. (2002). Aggregation of proteins with expanded glutamine and alanine repeats of the glutamine-rich and asparagine-rich domains of Sup35 and of the amyloid beta-peptide of amyloid plaques. Proc Natl Acad Sci USA *99*, 5596–5600.

Peters, R. (2005). Translocation Through the Nuclear Pore Complex: Selectivity and Speed by Reduction-of-Dimensionality. Traffic *6*, 421–427.

Peters, R. (2009). Translocation through the nuclear pore: Kaps pave the way. BioEssays *31*, 466–477.

Petri, M., Frey, S., Menzel, A., Goerlich, D., and Techert, S.A. (2012). Structural Characterization of Nanoscale Meshworks within a Nucleoporin FG Hydrogel. Biomacromolecules.

Plattner, H., and Klauke, N. (2001). Calcium in ciliated protozoa: sources, regulation, and calcium-regulated cell functions. Int. Rev. Cytol. *201*, 115–208.

Pollard, V.W., Michael, W.M., Nakielny, S., Siomi, M.C., Wang, F., and Dreyfuss, G. (1996). A novel receptor-mediated nuclear protein import pathway. Cell *86*, 985–994.

Powers, M.A., Macaulay, C., Masiarz, F.R., and Forbes, D.J. (1995). Reconstituted nuclei depleted of a vertebrate GLFG nuclear pore protein, p97, import but are defective in nuclear growth and replication. J Cell Biol *128*, 721–736.

Radu, A., Moore, M.S., and Blobel, G. (1995). The peptide repeat domain of nucleoporin Nup98 functions as a docking site in transport across the nuclear pore complex. Cell *81*, 215–222.

Ratner, G.A., Hodel, A.E., and Powers, M.A. (2007). Molecular determinants of binding between Gly-Leu-Phe-Gly nucleoporins and the nuclear pore complex. J Biol Chem *282*, 33968–33976.

Reichelt, R., Holzenburg, A., Buhle, E.L., Jarnik, M., Engel, A., and Aebi, U. (1990). Correlation between structure and mass distribution of the nuclear pore complex and of distinct pore complex components. J Cell Biol *110*, 883–894.

Rexach, M., and Blobel, G. (1995). Protein import into nuclei: association and dissociation reactions involving transport substrate, transport factors, and nucleoporins. Cell *83*, 683–692.

Ribbeck, K., and Görlich, D. (2001). Kinetic analysis of translocation through nuclear pore complexes. Embo J. *20*, 1320–1330.

Ribbeck, K., and Görlich, D. (2002). The permeability barrier of nuclear pore complexes appears to operate via hydrophobic exclusion. Embo J. *21*, 2664–2671.

Ribbeck, K., Kutay, U., Paraskeva, E., and Görlich, D. (1999). The translocation of transportin–cargo complexes through nuclear pores is independent of both Ran and energy. Current Biology *9*, 47–50.

Ribbeck, K., Lipowsky, G., Kent, H.M., Stewart, M., and Görlich, D. (1998). NTF2 mediates nuclear import of Ran. Embo J. *17*, 6587–6598.

Rout, M.P., Aitchison, J.D., Magnasco, M.O., and Chait, B.T. (2003). Virtual gating and nuclear transport: the hole picture. Trends in Cell Biology *13*, 622–628.

Rout, M.P., Aitchison, J.D., Suprapto, A., Hjertaas, K., Zhao, Y., and Chait, B.T. (2000). The yeast nuclear pore complex: composition, architecture, and transport mechanism. J Cell Biol *148*, 635–651.

Rout, M.P., and Blobel, G. (1993). Isolation of the Yeast Nuclear Pore Complex . J Cell Biol *123*, 771–783.

Salas-Marco, J., and Bedwell, D.M. (2004). GTP hydrolysis by eRF3 facilitates stop codon decoding during eukaryotic translation termination. Mol Cell Biol *24*, 7769–7778.

Sambrook, J.R. (2001). Molecular Cloning: A Laboratory Manual (Cold Spring Harbor Laboratory Press).

Santarella-Mellwig, R., Franke, J., Jaedicke, A., Gorjanacz, M., Bauer, U., Budd, A., Mattaj, I.W., and Devos, D.P. (2010). The compartmentalized bacteria of the planctomycetes-verrucomicrobia-chlamydiae superphylum have membrane coat-like proteins. PLoS Biol *8*, e1000281.

Sarma, A., and Yang, W. (2011). Calcium regulation of nucleocytoplasmic transport. Protein Cell *2*, 291–302.

Scheffzek, K., Klebe, C., Fritz-Wolf, K., Kabsch, W., and Wittinghofer, A. (1995). Crystal structure of the nuclear Ras-related protein Ran in its GDP-bound form. Nature *374*, 378–381.

Schmidt, C., and Urlaub, H. (2009). iTRAQ-labeling of in-gel digested proteins for relative quantification. Methods Mol Biol *564*, 207–226.

Schrader, N., Stelter, P., Flemming, D., Kunze, R., Hurt, E., and Vetter, I.R. (2008). Structural basis of the nic96 subcomplex organization in the nuclear pore channel. Mol Cell *29*, 46–55.

Schwoebel, E.D., Talcott, B., Cushman, I., and Moore, M.S. (1998). Ran-dependent signal-mediated nuclear import does not require GTP hydrolysis by Ran. J Biol Chem *273*, 35170–35175.

Shen, X., Yu, L., Weir, J.W., and Gorovsky, M.A. (1995). Linker histones are not essential and affect chromatin condensation in vivo. Cell *82*, 47–56.

Shevchenko, A., Wilm, M., Vorm, O., Jensen, O.N., Podtelejnikov, A.V., Neubauer, G., Shevchenko, A., Mortensen, P., and Mann, M. (1996). A strategy for identifying gel-separated proteins in sequence databases by MS alone. Biochem Soc Trans *24*, 893–896.

Shewmaker, F., Kryndushkin, D., Chen, B., Tycko, R., and Wickner, R. (2009). Two prion variants of Sup35p have in-register parallel -sheet structures, independent of hydration. Biochemistry.

Shewmaker, F., McGlinchey, R.P., and Wickner, R.B. (2011). Structural Insights into Functional and Pathological Amyloid. J Biol Chem *286*, 16533–16540.

Si, K., Choi, Y.-B., White-Grindley, E., Majumdar, A., and Kandel, E.R. (2010). Aplysia CPEB can form prion-like multimers in sensory neurons that contribute to long-term facilitation. Cell *140*, 421–435.

Simpson, A.G.B., and Roger, A.J. (2004). The real "kingdoms" of eukaryotes. Curr Biol *14*, R693–R696.

Siniossoglou, S., Lutzmann, M., Santos-Rosa, H., Leonard, K., Mueller, S., Aebi, U., and Hurt, E. (2000). Structure and assembly of the Nup84p complex. J Cell Biol *149*, 41–54.

Smith, A., Brownawell, A., and Macara, I.G. (1998). Nuclear import of Ran is mediated by the transport factor NTF2. Curr Biol *8*, 1403–1406.

Solmaz, S.R., Chauhan, R., Blobel, G., and Melčák, I. (2011). Molecular Architecture of the Transport Channel of the Nuclear Pore Complex. Cell *147*, 590–602.

Stansfield, I., Jones, K.M., Kushnirov, V.V., Dagkesamanskaya, A.R., Poznyakovski, A.I., Paushkin, S.V., Nierras, C.R., Cox, B.S., Ter-Avanesyan, M.D., and Tuite, M.F. (1995). The products of the SUP45 (eRF1) and SUP35 genes interact to mediate translation termination in Saccharomyces cerevisiae. Embo J. *14*, 4365–4373.

Stavru, F., Hülsmann, B.B., Spang, A., Hartmann, E., Cordes, V.C., and Görlich, D. (2006). NDC1: a crucial membrane-integral nucleoporin of metazoan nuclear pore complexes. J Cell Biol *173*, 509–519.

Stewart, M., Kent, H.M., and McCoy, A.J. (1998). Structural basis for molecular recognition between nuclear transport factor 2 (NTF2) and the GDP-bound form of the Ras-family GTPase Ran. J Mol Biol *277*, 635–646.

Stoffler, D., Feja, B., Fahrenkrog, B., Walz, J., Typke, D., and Aebi, U. (2003). Cryo-electron tomography provides novel insights into nuclear pore architecture: implications for nucleocytoplasmic transport. J Mol Biol *328*, 119–130.

Strawn, L.A., Shen, T., Shulga, N., Goldfarb, D.S., and Wente, S.R. (2004). Minimal nuclear pore complexes define FG repeat domains essential for transport. Nature Cell Biology *6*, 197–206.

Sukegawa, J., and Blobel, G. (1993). A nuclear pore complex protein that contains zinc finger motifs, binds DNA, and faces the nucleoplasm. Cell *72*, 29–38.

Sunde, M., and Blake, C. (1997). The structure of amyloid fibrils by electron microscopy and X-ray diffraction. Adv. Protein Chem. *50*, 123–159.

Tamura, K., Fukao, Y., Iwamoto, M., Haraguchi, T., and Hara-Nishimura, I. (2010). Identification and Characterization of Nuclear Pore Complex Components in Arabidopsis thaliana. The Plant Cell Online *22*, 4084–4097.

Tetenbaum-Novatt, J., Hough, L.E., Mironska, R., McKenney, A.S., and Rout, M.P. (2012). Nucleocytoplasmic transport: A role for non-specific competition in karyopherin-nucleoporin interactions. Mol Cell Proteomics.

Toombs, J.A., McCarty, B.R., and Ross, E.D. (2010). Compositional determinants of prion formation in yeast. Mol Cell Biol *30*, 319–332.

Tourancheau, A.B., Tsao, N., Klobutcher, L.A., Pearlman, R.E., and Adoutte, A. (1995). Genetic code deviations in the ciliates: evidence for multiple and independent events. Embo J. *14*, 3262–3267.

True, H.L., and Lindquist, S.L. (2000). A yeast prion provides a mechanism for genetic variation and phenotypic diversity. Nature *407*, 477–483.

Unwin, P.N., and Milligan, R.A. (1982). A large particle associated with the perimeter of the nuclear pore complex. J Cell Biol *93*, 63–75.

van der Wel, P.C.A., Lewandowski, J.R., and Griffin, R.G. (2007). Solid-state NMR study of amyloid nanocrystals and fibrils formed by the peptide GNNQQNY from yeast prion protein Sup35p. J Am Chem Soc *129*, 5117–5130.

Vetter, I.R., Arndt, A., Kutay, U., Görlich, D., and Wittinghofer, A. (1999). Structural view of the Ran-Importin beta interaction at 2.3 A resolution. Cell *97*, 635–646.

Voet, D., and Voet, J.G. (2010). Biochemistry (Wiley).

Watson, M.L. (1954). Pores in the mammalian nuclear membrane. Biochim Biophys Acta *15*, 475–479.

Weathers, E.A., Paulaitis, M.E., Woolf, T.B., and Hoh, J.H. (2004). Reduced amino acid alphabet is sufficient to accurately recognize intrinsically disordered protein. FEBS Lett *576*, 348–352.

Weis, K. (2007). The nuclear pore complex: oily spaghetti or gummy bear? Cell *130*, 405–407.

White, E.M., Allis, C.D., Goldfarb, D.S., Srivastva, A., Weir, J.W., and Gorovsky, M.A. (1989). Nucleus-specific and temporally restricted localization of proteins in Tetrahymena macronuclei and micronuclei. J Cell Biol *109*, 1983–1992.

Whittle, J.R.R., and Schwartz, T.U. (2009). Architectural nucleoporins Nup157/170 and Nup133 are structurally related and descend from a second ancestral element. J Biol Chem *284*, 28442–28452.

Wischnitzer, S. (1958). An electron microscope study of the nuclear envelope of amphibian oocytes. J. Ultrastruct. Res. *1*, 201–222.

Wu, M., Allis, C.D., Sweet, M.T., Cook, R.G., Thatcher, T.H., and Gorovsky, M.A. (1994). Four distinct and unusual linker proteins in a mitotically dividing nucleus are derived from a 71-kilodalton polyprotein, lack p34cdc2 sites, and contain protein kinase A sites. Mol Cell Biol *14*, 10–20.

Yamada, J., Phillips, J.L., Patel, S., Goldfien, G., Calestagne-Morelli, A., Huang, H., Reza, R., Acheson, J., Krishnan, V.V., Newsam, S., et al. (2010). A bimodal distribution of two distinct categories of intrinsically-disordered structures with separate functions in FG nucleoporins. Mol Cell Proteomics.

Yang, Q., Rout, M.P., and Akey, C.W. (1998). Three-dimensional architecture of the isolated yeast nuclear pore complex: functional and evolutionary implications. Mol Cell *1*, 223–234.

Zhouravleva, G., Frolova, L., Le Goff, X., Le Guellec, R., Inge-Vechtomov, S., Kisselev, L., and Philippe, M. (1995). Termination of translation in eukaryotes is governed by two interacting polypeptide chain release factors, eRF1 and eRF3. Embo J. *14*, 4065–4072.

Acknowledgments

As Christian de Duve said, 'good research is not learned in books, but at the bench, like the crafts in the middle ages, under the supervision of a master'. Hence, I first and foremost want to thank my supervisor, Prof. Dr. Dirk Görlich, for having shared his mastery of science with me. Dirk, I am grateful for all your input to this work, but even more for what you have taught me beyond that. Thank you especially for constantly pushing me for more conceptual clarity.

I also want to express my gratitude to my thesis committee members Prof. Dr. Peter Rehling and Prof. Dr. Helmut Grubmüller for their support and advice. Likewise, I want to thank Prof. Dr. Blanche Schwappach, Prof. Dr Silvio Rizzoli and Prof. Dr. Detlef Doenecke for serving on my extended committee. It is great to have all the scientists that accompanied me throughout the last years coming together for my thesis defense.

A special thanks goes to Blanche, who has supported me ever since my first molecular biology lecture and is an invaluable mentor. In this regard, I also like to thank Prof. Dr. Mary Osborn and Prof. Dr. Reinhard Jahn.

Science oftentimes is a collaborative effort. Prof. Dr. Henning Urlaub, Monika Raabe, Lisa Neuenroth and Uwe Plessmann helped me to identify the various proteins that I 'fished' out of the *Tetrahymena* proteome with their mass spectrometry analysis. Dr. Peter Odenwälder supplied me with 'moderate' amounts of HeLa extracts, as he called it. Prof. Dr. Markus Engstler and Dr. Nicola Jones contributed *Trypanosoma* DNA. Dr. Steffen Frey shared his extensive plasmid database, permeation probes and Nsp1 FG repeats. Last, but not least, Tino Pleiner provided the biotinylated $Imp\beta_{45-462}$ mutant that helped identifying even more curious FG repeat-like proteins. Thanks to you all.

This work would not have been possible without the help of Jürgen Schünemann, who performed all the countless HPLC runs of the FG repeats purified in this study. Susanne Brandfass helped with the cloning. Gabrielle Kopp prepped lots and lots of DNA for me. Uwe Hoffmann, Elisabeth Richert and Gabriele Kopp handled and organized the chaos of the wet lab and Cornelia Paz of the dry lab. I am grateful to all of you for your help.

Science however is also collaborative in another way: many ideas are only born when sharing experience and excitement with your colleagues. In this regard, I especially like to thank Dr. Steffen Frey, Koray Kirli, Aksana Labohka, Michael Ridders, Dr. Matthias Samwer, Dr. Thomas Güttler, Dr. Volker Cordes and Dr. Annette Denker. Moreover, thanks to everyone else from the Görlich lab. Of course, also the MolBio retreats and seminars have always been great opportunities to discuss science. Thanks to everyone who contributed.

In biology, 'structure is function'. I want to express my gratitude to all the people involved in the Molecular Biology program and the GGNB for providing such a great infrastructure for me to 'function', especially Dr. Steffen Burkhardt and Kirsten Pöhlker. Thank you also for your support to make Horizons and WoCaNet a success. I am also grateful to the Max Planck Society for generously supporting me during the last years.

At the end, I want to thank my family and friends for supporting me. Knowing me, I am afraid that it oftentimes was rather an enduring me. Hence, all the more, thanks Mama, Papa, Henry, Torge, Verena, Fred, Cornelius, Jenny, Micha, Tino, Jens, Christoph, Anika, Erik, Peter, Max, Clemens, Marc – and of course mi novia maravillosa Anne.

Curriculum vitae

Broder Schmidt
Molecular Cell Biologist, BSc
Born: April 30th, 1985 in Itzehoe, Germany

Current adress:
Max Planck Institute for Biophysical Chemistry
Karl-Friedrich-Bonhoeffer Institute
Am Faßberg 11
37077 Göttingen
Phone: +49-551-2012413
Mail: hb.schmidt@mpibpc.mpg.de

Education	2008 – 2012	PhD thesis in the group of Prof. Dr. Dirk Görlich at the Max Planck Institute for Biophysical Chemistry.	Göttingen
	2008 – 2012	Göttingen Graduate School for Neurosciences, Biophysics and Molecular Biosciences (GGNB)	Göttingen
	2008	Completion of practical and theoretical MSc studies.	Göttingen
	2007 – present	MSc/PhD Program in Molecular Biology (International Max Planck Research School, University of Göttingen)	Göttingen
	2004 – 2007	Studies in Molecular and Cellular Biology (University of Heidelberg, Award of the degree 'Bachelor of Science)	Heidelberg
	2002 – 2004	Bismarck Gymnasium	Elmshorn
	2001 – 2002	Oak Hills High School	Cincinnati
	1997 – 2001	Bismarck Gymnasium	Elmshorn
	1995 - 1997	Realschule am Probstenfeld	Elmshorn

Peer-reviewed Publications	Christian Ader, Steffen Frey, Werner Maas, **Hermann Broder Schmidt**, Dirk Görlich and Marc Baldus (2010). Amyloid-like interactions within nucleoporin FG hydrogels. *Proc. Natl. Acad. Sci.* Vol. 107 pp. 6281-6285.
	Frederik Köpper, Birgit Manno and **Broder Schmidt** (2010). Wovon wir leben ist, wovon wir sterben - Von der Suche nach den molekularbiologischen Grundlagen für Leben und Tod (Scientific Essay). *Journal 360°*. Issue 1, pp. 16-24.

Research Experience and Internships	2008 – 2012	PhD thesis in the group of Prof. Dr. Dirk Görlich at the Max Planck Institute for Biophysical Chemistry on: *Requirements of FG repeats to form functional hydrogel-based NPC-like permeability barriers.*
	06/2008 – 07/2008	Internship with Prof. Dr. Dirk Görlich, Max Planck Institute for Biophysical Chemistry: *The Distance Between Nuclear Transport Receptor and Cargo Influences Nuclear Import, But Not Enrichement In FG-Repeat Hydrogels.* and *The TAP Nuclear Localisation Signal Mediates Efficient Cargo Enrichment in the Nuclei of Permeabilized Cells and FG-Repeat Hydrogels.*
	03/2008 – 04/2008	Internship with Prof. Dr. Ralf Ficner, Göttingen Center for Molecular Biosciences, University of Göttingen: *Cloning, Purification and Crystallization of the Eukaryotic Translation Initiation Factor eIF2α.*
	01/2008 – 02/2009	Internship with Prof. Dr. Reinhard Jahn, Max Planck Institute for Biophysical Chemistry: *A Novel Approach to Study the Homotypic Fusion of Early Endosomes.*
	07/2007 – 08/2007	Internship with Prof. Dr. Irmgard Sinning, Biochemistry Center of the University of Heidelberg: *Cloning, Purification and Crystallization of the G-domain of the chloroplast translocon subunit Toc159."*
	03/2007 – 05/2007	BSc thesis with Prof. Dr. Blanche Schwappach and Prof. Dr. Elmar Schiebel, Center for Molecular Biology of the University of Heidelberg: *Analysis of the Cell Cycle in GET-gene mutant yeast.*
Extracurricular Engagement	2011 – 2012	Student representative on the executive board of the Göttingen Graduate School for Neurosciences, Biophysics and Molecular Biosciences (GGNB).
	2010 – 2011	Initiation and organization of the *Women's Careers and Networks* symposium (www.wocanet.uni-goettingen.de)
	2009 – 2011	Organization of the annual international PhD student symposium *Horizons in Molecular Biology* (www.horizons.uni-goettingen.de)
	2009	Organization of the 1st e-fellows.net Charity Run.